International Conference on Intelligent and Sustainable Power and Energy Systems (ICISPES 2023)

IOP Conference Series: Materials Science and Engineering
Volume 1295

Bangalore, India
26 - 27 September 2023

ISBN: 978-1-7138-8732-4
ISSN: 1757-8981

Printed from e-media with permission by:

Curran Associates, Inc.
57 Morehouse Lane
Red Hook, NY 12571

Some format issues inherent in the e-media version may also appear in this print version.

Institute of Physics
Dirac House, Temple Back
Bristol BS1 6BE UK

Phone: 44 1 17 929 7481
Fax: 44 1 17 920 0979

techtracking@iop.org

Additional copies of this publication are available from:

Curran Associates, Inc.
57 Morehouse Lane
Red Hook, NY 12571 USA
Phone: 845-758-0400
Fax: 845-758-2633
Email: curran@proceedings.com
Web: www.proceedings.com

TABLE OF CONTENTS

Proceedings

1st International Conference on Intelligent and Sustainable Power and Energy Systems

(ICISPES 2023)

26 - 27 September, 2023

Publication Partner

IOP Publishing
Conference Series
Materials Science and Engineering

Financial Partner

E-LAGORi
Prototype Your Dreams

Organised by

Department of Electrical and Electronics Engineering
Dayananda Sagar College of Engineering
Kumaraswamy Layout, Bangalore, Karnataka, India
www.dsce.edu.in

IOP Conf. Series: Materials Science and Engineering 1295 (2023) 011001 doi:10.1088/1757-899X/1295/1/011001

About the Conference

The "**1st International Conference on Intelligent and Sustainable Power and Energy Systems (ICISPES 2023)**" is aimed to address the state-of-the-art multidisciplinary research needs and interdisciplinary aspects of intelligent and sustainable technologies in power and energy systems in the form of research papers from industry, faculty, research scholars, and PG/UG students, along with keynote lectures and a number of invited talks from reputed speakers. The ICISPES 2023 aims to provide a platform to present their work to share experiences and ideas in the areas of Renewable Energy, Electric Transportation, Power Electronics, Electric Drives, Control Techniques, Smart Grids, Communication Protocols, Intelligent Charging Infrastructure and Standards, machine learning, artificial intelligence, big data analytics, IoT, etc. In view of the changing scenario under intelligent and sustainable power and energy, this conference aims to put together the experts from these areas to disseminate their knowledge and experience for working in years to come.

The ICISPES provides an international platform for all research scholars, practitioners, academicians, students and scientists to discuss and exhibit the latest research developments and discoveries in intelligent and sustainable technologies. This conference enables us to share and exchange research knowledge and ideas in power and energy systems.

IOP Conf. Series: Materials Science and Engineering 1295 (2023) 011001 doi:10.1088/1757-899X/1295/1/011001

About the Institution

Dayananda Sagar College of Engineering is approved by the All-India Council for Technical Education (AICTE), Govt. of India and affiliated to Visvesvaraya Technological University, Belgaum, Karnataka. It has the widest choice of Engineering branches, having 20 undergraduate courses and 06 postgraduate courses. In addition, it has 20 research centres in different branches of Engineering catering to research scholars for obtaining a PhD under VTU. Various courses are accredited by the NBA. The Institute is spread over 23 acres of land with a requisite infrastructure supported by laboratories with state-of-the-art equipment and machines. The central library, with modern facilities and a digital library, provides the knowledge base for the students. The college has good faculty strength with the highest professional integrity and commitment to academics with transparency in their actions. The faculty members are research-oriented, having a number of sponsored R&D projects from different agencies such as the Department of Science & Technology, Defense R&D organizations, Indian Space Research Organization, AICTE, etc.

About the Department

Established in 1979, the department started the UG Program with an intake of 50, and later, the intake was increased to 120 in the year 2006. The department launched an M.Tech. program in Power Electronics in the year 2006. Further, the EEE Department is recognized as an R&D Centre by VTU, which offers doctoral programs- Ph.D. and M.Sc. (Engg.) by research program. The department is approved by AICTE and has been accredited four times (the last accreditation in Tier-1 from 2022 to 2025) by NBA, New Delhi.

Preface

The unrelenting quest for an enhanced, healthier, and more comfortable existence has spurred innovation. This technological evolution and its implementation in practical solutions is a pivotal force propelling contemporary society forward. Current advancements in various engineering fields, including electrical, electronics, mechanical, communications, chemical, and computer fields such as AI, target developing more intelligent systems encapsulated in concepts like smart homes, smart cities, and smart societies. Furthermore, innovation in renewable energy, electric transportation, power electronics, electric drives, control techniques, smart grids, sustainability, communication protocols, and intelligent charging infrastructures and standards are intrinsically interconnected. Innovations in one sphere address specific issues and potentially facilitate solutions or foster advancements in other domains. Therefore, ongoing research in these sectors must be paired with effectively disseminating progress and findings. This approach is crucial for fostering a global society that is continuously evolving and advancing towards a future that embraces the synergistic potential of these innovations.

The International Conference on Intelligent and Sustainable Power and Energy Systems, also known as ICISPES 2023, is set to be hosted by the Department of Electrical and Electronics Engineering at the Dayananda Sagar College of Engineering, Bangalore, Karnataka, India, from September 26-27, 2023. This pivotal event is poised to serve as a vibrant hub for the convergence of groundbreaking ideas from researchers worldwide, steering our cities and societies towards greater intelligence and integration. The ICISPES 2023 is dedicated to fostering innovation and collaboration across diverse yet interconnected engineering and technology domains, acknowledging that each field contributes vital to the betterment of society. To this end, the call for papers was crafted to encourage a substantial pool of contributions, attracting over 48 online submissions for IOP publication from various relevant fields. Out of this rich array of scholarly contributions, only 18 high-calibre submissions have been chosen to feature in the conference, a selection made possible through a stringent and impartial peer-review process. This review system encompassed the expertise and insights of a distinguished panel of experienced researchers, advisory members, and the technical and program committee of ICISPES 2023. The selection criteria were grounded in the significance of the contributions made by the authors, as well as the technical depth, originality, and alignment with the conference's central themes. It is worth noting that every stage of the peer-review process was facilitated seamlessly through an electronic system, ensuring efficiency and transparency.

The keynote address is a pivotal segment in any conference, facilitating the dissemination of research findings and insights before an audience of scientists and researchers from various disciplines and backgrounds. Acknowledging this, the organizers of ICISPES 2023 have exerted considerable effort to coordinate a series of keynote speeches that align perfectly with the conference's thematic focus. Furthermore, the assembly of author presentations has been meticulously orchestrated to highlight the significance and interconnectedness of the different pieces while resonating with the foundational philosophy and driving forces behind the conference. In light of this, we extend our heartfelt appreciation to the authors for choosing 'ICISPES' as the platform to share the fruits of their research endeavours. Moreover, we are profoundly thankful for the invaluable assistance rendered by each individual reviewer and the members of the Program Committee, whose contributions to the peer-review process have been instrumental in elevating the quality and substance of the conference.

Acknowledgement

The editors are honoured to have facilitated the presentation of pivotal proposals and highlights of the ICISPES 2023 Conference. This event has garnered the attention and enthusiasm of academics and researchers worldwide, establishing itself as an optimal platform to unveil a myriad of innovative research discoveries. The submissions spanned various disciplines, including electrical, electronics, computation, and agricultural engineering.

We wish to express our heartfelt appreciation to all the authors who have amplified the value of the conference with their precious contributions, dedicating their time, expertise, and insightful research submissions. The seamless execution of pre- and post-conference activities and deliberations was made possible through our national and international advisory committees' timely and informed support and guidance. We convey our profound gratitude to all committee members for their invaluable input. Furthermore, we would like to acknowledge the remarkable efforts of our robust team of reviewers, who conducted thorough and critical evaluations of all submissions, offering feedback and suggestions that were vital in maintaining the high standard and quality of the conference proceedings.

A special note of thanks goes to all organising committee members, whose relentless efforts were pivotal in making this event a resounding success. We also extend our deepest gratitude to the editorial team at IOP Conference Series: Materials Science and Engineering for the exceptional work in crafting the proceedings and journal in an innovative and intellectually stimulating manner. The ICISPES 2023 conference and its proceedings stand to receive well-deserved accolades in front of a vast assembly of attendees.

Finally, our heartfelt thanks to the Management of Dayananda Sagar College of Engineering, Karnataka, India and the Department of Electrical and Electronics Engineering faculty members for their continued support and encouragement to make the conference a huge success.

IOP Conf. Series: Materials Science and Engineering 1295 (2023) 011001 doi:10.1088/1757-899X/1295/1/011001

Message from the Authorities

IOP Conf. Series: Materials Science and Engineering 1295 (2023) 011001 doi:10.1088/1757-899X/1295/1/011001

Dr. D. Hemachandra Sagar, M.B.B.S., M.S.

Chairman, DSI

Message

I am happy to know that the Electrical and Electronics Engineering department is organizing the 1st **International Conference on Intelligent and Sustainable Power and Energy Systems (ICISPES 2023)** on 26th and 27th September 2023 in **Dayananda Sagar College of Engineering**, Bangalore, India.

In the present-day scenario, the development of innovative sustainable technologies for power and energy systems holds significant importance. Establishing an international forum to exchange ideas and showcase the most recent research breakthroughs and findings has the potential to pave the way for sustainable progress. I also believe our students will immensely benefit from this forum as it will help them become industry-ready.

I wish the International Conference 'ICISPES 2023' a grand success.

Dr. D. Hemachandra Sagar, M.B.B.S., M.S.

Chairman, DSI

IOP Conf. Series: Materials Science and Engineering 1295 (2023) 011001 doi:10.1088/1757-899X/1295/1/011001

Dr. D. Premachandra Sagar

Vice Chairman, DSI

Message

Dayananda Sagar Institutions have consistently been at the forefront of enhancing and fostering an educational system enriched with a culture of research. I'm delighted to learn that the Department of Electrical and Electronics Engineering is hosting the International Conference on Intelligent and Sustainable Power and Energy Systems (ICISPES 2023) at Dayananda Sagar College of Engineering in Bangalore, India, on September 26th and 27th, 2023.

ICISPES 2023 offers a global forum for research scholars, practitioners, academicians, students, and scientists to engage in discussions and showcase the most recent research advancements and findings in the realm of intelligent and sustainable technologies. This is a pertinent and highly essential area of concentration at this moment. I hope this forum leads to new collaborations with various research scholars and industry leaders for the latest innovations.

I extend a warm welcome to all the delegates and wish the event a great success.

Dr. D. Premachandra Sagar

Vice Chairman, DSI

IOP Conf. Series: Materials Science and Engineering 1295 (2023) 011001 doi:10.1088/1757-899X/1295/1/011001

Shri Galiswamy

Secretary, DSI

Message

I am pleased to learn that the Electrical and Electronics Engineering department of DSCE is organizing the 1st International Conference on Intelligent and Sustainable Power and Energy Systems (ICISPES 2023) on 26th and 27th September 2023 in Dayananda Sagar College of Engineering, Bangalore, India.

DSCE stands as a premier educational institution known for its regular hosting of conferences, and this conference will provide an opportunity for researchers and young minds to collaborate and share their knowledge on a common platform, thereby benefiting all involved. Through these forums we aim to inspire and encourage our students to reach greater heights in their careers.

I extend my best wishes for the great success of the International Conference 'ICISPES 2023.'

Shri Galiswamy

Secretary, DSI

IOP Conf. Series: Materials Science and Engineering 1295 (2023) 011001 doi:10.1088/1757-899X/1295/1/011001

Dr. B.G. Prasad

Principal, DSCE

Message

It is remarkable to witness the 1st International Conference on 'Intelligent and Sustainable Power and Energy Systems (ICISPES 2023)' by the E&EE department on 26th and 27th September 2023 in Dayananda Sagar College of Engineering, Bangalore, India.

This conference is focused on Power and Energy Systems and is crucial as it facilitates the progression of knowledge, encourages innovation, addresses pressing energy challenges, and encourages collaboration among various stakeholders. This initiative plays a central role in propelling the development of sustainable and efficient energy systems. I encourage all the delegates to make the most of this opportunity and interact with one another to get enriched with the latest technological developments.

I applaud the organizing committee for their dedicated efforts and extend my best wishes for a resounding success to all the participating delegates and research scholars.

Dr. B.G. Prasad

Principal, DSCE

IOP Conf. Series: Materials Science and Engineering 1295 (2023) 011001 doi:10.1088/1757-899X/1295/1/011001

Dr. P. Usha

Organizing Chair

ICISPES 2023

Message

The international conference 'Intelligent and Sustainable Power and Energy Systems (ICISPES 2023)' offers a global forum for research scholars, practitioners, academicians, students, and scientists to engage in discussions and showcase the most recent research advancements and findings in the realm of intelligent and sustainable technologies.

I would like to express my appreciation to all the esteemed delegates of the international conference 'Intelligent and Sustainable Power and Energy Systems (ICISPES 2023)' who have enriched and elevated this event through their participation and active engagement.

I also would like to thank the organizing committee (my faculty and staff) who made this event possible with their time and effort. We in the E&EE department continuously strive to create a better learning environment for our students.

I have confidence that this conference will produce valuable results that will benefit all participants, and we hope to organize more such events in future.

Dr. P. Usha

Organizing Chair

ICISPES 2023

IOP Conf. Series: Materials Science and Engineering 1295 (2023) 011001 doi:10.1088/1757-899X/1295/1/011001

Speakers

Prof. Emeritus. Akira Mizuno

Toyohashi University of Technology, Japan

Prof. Kameswara Musti

Associate Professor, Electrical and Computer Engineering, Namibia University of Science and Technology, Namibia

Prof. Dr. Ramesh Rayudu

Associate Professor, Deputy Head of School, School of Engineering and Computer Science, Victoria University of Wellington, Wellington, New Zealand

Organizing Committee

Chief Patrons

Chairman
Dr. D Hemachandra Sagar
Dayananda Sagar Institutions
Bengaluru

Vice-Chairman
Dr. D Premachandra Sagar
Dayananda Sagar Institutions
Bengaluru

Patrons

Secretary
Shri. Galiswamy
Dayananda Sagar Institutions
Bengaluru

Joint Secretary
Ms. Tintisha H Sagar
Dayananda Sagar Institutions
Bengaluru

Joint Secretary
Mr. Rohan Prem Sagar
Dayananda Sagar Institutions
Bengaluru

Joint Secretary
Mr. Nishan H Sagar
Dayananda Sagar Institutions
Bengaluru

Joint Secretary
Ms. Natasha Sagar
Dayananda Sagar Institutions
Bengaluru

Honorary Chair

Dr. H P Khincha
Advisor
Dayananda Sagar Institutions, Bengaluru

Organising Chair

Principal
Dr. B G Prasad
Dayananda Sagar College of Engineering
Bengaluru

Professor and Head
Dr. P Usha
Department of Electrical and Electronics Engineering
Dayananda Sagar College of Engineering
Bengaluru

Local Advisory Committee

Vice Principal
Dr. Ramesh Babu D R
Dayananda Sagar College of Engineering, Bengaluru

Vice Principal
Dr. Ramaraju H K
Dayananda Sagar College of Engineering, Bengaluru

Vice Principal
Dr. Suma V
Dayananda Sagar College of Engineering, Bengaluru

Dean-Academics
Dr. A R Aswatha
Dayananda Sagar College of Engineering, Bengaluru

Professor
Dr. Abraham George
Department of Electrical and Electronics Engineering
Dayananda Sagar College of Engineering, Bengaluru

Professor
Dr. A D Srinivasan
Department of Electrical and Electronics Engineering
Dayananda Sagar College of Engineering, Bengaluru

International Technical Advisory Committee

- **Dr. Abhishek Pandey**, Electrical, Instrumentation, and Controls Engineer, Water and Wastewater Systems, Houston, Texas, United States.

- **Mr. Sanjay N**, Embedded Technical Lead at Mirion Technologies (Capintec), Inc., New York City Metropolitan Area.
- **Dr. Fernando Yanine**, Faculty of Engineering, Universidad Finis Terrae, Santiago, Chile.
- **Dr. Krzysztof Lalik**, University of Science and Technology AGH, Kraków, Poland.
- **Dr. Hassan Haes Alhelou**, Chief Investigator, Department of Electrical and Computer Systems Engineering, Monash University, Clayton VIC 3800, Australia.
- **Prof. Akira Mizuno**, Senior Researcher, Professor Emeritus, Department of Environmental and Life Sciences, Toyohashi University of Technology, Japan.
- **Dr. Noor Izzri bin Abdul Wahab**, Associate Professor, Advanced Lightning, Power and Energy Research (ALPER), Dept. of Electrical and Electronic Engineering, Universiti Putra Malaysia.
- **Mr. Shashank B L**, Senior System Architect, Micron Technology, Folsom, California, USA.
- **Prof. Kameswara Musti**, Associate Professor, Electrical and Computer Engineering, Namibia University of Science and Technology, Namibia.
- **Dr. Ramesh Rayudu**, Associate Professor, Deputy Head of School, School of Engineering and Computer Science, Victoria University of Wellington, Wellington, New Zealand.

National Technical Advisory Committee

- **Dr. Chandrasekhar Reddy Atla**, DGM, PRDC, Bengaluru.
- **Dr. Anand R**, GM, Otis, Bengaluru.
- **Dr. Sukumar Mishra**, Prof. & Assoc. Dean R&D, IIT, Delhi.
- **Dr. R Sarathi**, IIT Madras.
- **Dr. Rajanikanth**, B.S., IISC, Bengaluru.
- **Dr. Rathna G N**, Associate Professor, Principal Research Scientist, Systems Science And Signal Processing, Indian Institute of Science, Bengaluru.
- **Dr. Subramanyam**, IIT, Hyderabad.
- **Dr. C. Vyjayanthi**, Assoc. Prof., Dept. of EE, NIT, Goa.
- **Dr. Tukaram Mogar**, Dept. of EEE, NITK, Suratkal.
- **Dr. Ashish Mani**, Prof., E&EE, Amity University.
- **Dr. Ravi K**, Prof., EEE, VIT Vellore.
- **Dr. Suresh Babu**, Assoc. Director, NPTI, Bengaluru.
- **Dr. Sarat Kumar Sahoo**, Dept. EE, Govt. Eng College, Berhampur, Odisha.
- **Dr. Sanjiba Bisoyi**, Assoc. Prof., JSS Academy of Technical Education, Noida.
- **Dr. Bhagwan Shree Ram**, Prof. & Head, Dept. of EEE, Saharsa College of Engineering, Bihar.
- **Dr. Vikramaditya Dave**, Assoc. Prof., Dept. of EEE, College of Technology and Engineering, Udaipur (CTAE) (Govt. University).

- **Dr. Kannan Tinnium**, Technology Director – External Programs, Energy Storage Center of Excellence, Secure Power, Bengaluru.

Department Coordinators

- **Dr. Dhanalakshmi R**, Professor, Dept. of EEE, DSCE
- **Dr. Premkumar M**, Associate Professor, Dept. of EEE, DSCE
- **Dr. Sharath Kumar Y N**, Assistant Professor, Dept. of EEE, DSCE

Publishing Committee

- **Dr. Premkumar M**, Associate Professor, Dept. of EEE, DSCE
- **Dr. Sujit Kumar**, Assistant Professor, Dept. of EEE, DSCE
- **Ms. Suganthi N**, Assistant Professor, Dept. of EEE, DSCE
- **Ms. Supriya J**, Assistant Professor, Dept. of EEE, DSCE

Review Committee

- **Dr. Abraham George**, Professor, Dept. of EEE, DSCE
- **Dr. Balamurugan M**, Assistant Professor, Dept. of EEE, DSCE
- **Ms. Shubha Kulkarni**, Assistant Professor, Dept. of EEE, DSCE
- **Ms. Sumana S**, Assistant Professor, Dept. of EEE, DSCE

Registration Committee

- **Dr. Soni M**, Assistant Professor, Dept. of EEE, DSCE
- **Ms. Rachana M Hullamani**, Assistant Professor, Dept. of EEE, DSCE
- **Ms. Deekshitha Arasa**, Assistant Professor, Dept. of EEE, DSCE
- **Ms. Shruthi G**, Assistant Professor, Dept. of EEE, DSCE

Sponsorship Committee

- **Dr. A D Srinivasan**, Professor, Dept. of EEE, DSCE
- **Dr. Arjun Kumar G B**, Assistant Professor, Dept. of EEE, DSCE
- **Mr. Satish B A**, Assistant Professor, Dept. of EEE, DSCE

Logistics and Hospitality Committee

- **Dr. Suchana Mishra**, Assistant Professor, Dept. of EEE, DSCE
- **Ms. Kavana Salimath**, Assistant Professor, Dept. of EEE, DSCE
- **Mr. Nandan N**, Assistant Professor, Dept. of EEE, DSCE
- **Mr. Ganesha G C**, Assistant Professor, Dept. of EEE, DSCE

Technical Support Committee

- **Dr. P B Savitha**, Associate Professor, Dept. of EEE, DSCE

IOP Conf. Series: Materials Science and Engineering 1295 (2023) 011001 doi:10.1088/1757-899X/1295/1/011001

- **Dr. Gopisetti Manikanta**, Assistant Professor, Dept. of EEE, DSCE
- **Ms. Tanuja H**, Assistant Professor, Dept. of EEE, DSCE
- **Ms. Sreevidya T R**, Assistant Professor, Dept. of EEE, DSCE
- **Ms. Shruti S Athanikar**, Assistant Professor, Dept. of EEE, DSCE
- **Ms. Kavitha K**, Assistant Professor, Dept. of EEE, DSCE

Treasurer

- **Dr. Sharath Kumar Y N**, Assistant Professor, Dept. of EEE, DSCE

Peer Review Statement

All papers published in this volume have been reviewed through processes administered by the Editors. Reviews were conducted by expert referees to the professional and scientific standards expected of a proceedings journal published by IOP Publishing.

- **Type of peer review:** Single Anonymous
- **Conference submission management system:** Morressier
- **Number of submissions received:** 48
- **Number of submissions sent for review:** 48
- **Number of submissions accepted:** 18
- **Acceptance Rate (Submissions Accepted / Submissions Received × 100):** 37.5
- **Average number of reviews per paper:** 0.99
- **Total number of reviewers involved:** 31
- **Contact person for queries:**
  Name: Usha P
  Email: hod-eee@dayanandasagar.edu
  Affiliation: Dayananda Sagar Institutions - Dayananda Sagar College of Engineering

IOP Conf. Series: Materials Science and Engineering 1295 (2023) 012001 doi:10.1088/1757-899X/1295/1/012001

Solar-Powered Peltier-Based Air Cooler and Warmer in Medical Systems

Arjun Kumar GB[1*], Balamurugan M[1], Sunil Kumar KN[3], Ravi Gatti[2], Ganesh P[1] and Tasnim Sallauddin Kotwal[1]

[1] Department of Electrical and Electronics Engineering, Dayananda Sagar College of Engineering, Bengaluru, Karnataka, India
[2] Department of Electronics & Communication Engineering, Dayananda Sagar Academy of Technology and Management, Bengaluru, Karnataka, India
[3] Department of Electrical and Electronics Engineering, Sri Venkateshwara College of Engineering, Bengaluru, Karnataka, India

arjun-eee@dayanandasagar.edu

Abstract. Presently, there are many more commercial applications for thermoelectric Peltier coolers, notably portable air/water refrigerators, electronics cooling processes, thermal management systems in medical applications, and others. Thus, the goal of this project is to develop an experimentally-based optimisation procedure for powered by sunlight peltier-based air conditioners and heaters for use in hospitals that are capable of running continuously. This enables patients to cook meals in the heating element and keep drugs in the cold compartment inside the cooler, both functions being incorporated into one model while also reducing the device's dimensions and requirement for space. In this study, a model that employs perturb and observe (P&O) approaches applied to solar systems under constant and partially shady circumstances was powered by the Maximum Power Point Tracking (MPPT) methodology. Additionally, the Internet of Things modules offer user and autonomous management, while the entire system is tracked and shown on an OLED. An appropriate number of modules also reduces the expense per cooling unit for Peltier coolers and warmers.

1. Introduction

The amount of energy used in our daily lives is rising as a result of the declining availability of sources of green energy like coal, petroleum, and other such resources. The heavy reliance on renewable energy sources like coal and oil for power causes numerous ecological issues, such as pollution. With the help of renewable energy sources like wind and solar power, we can address issues with the atmosphere's ozone layer, climate change, and electricity consumption. Our goal is to present a portable refrigeration unit that uses a Peltier module and the sun's rays as its power source to get around the drawbacks of the current model. As the global population grows and the natural environment degrades at a staggering pace, thermoelectric effect couple systems have emerged as a viable alternative because they are reasonably priced, sustainable, and small in dimensions [1].

Worldwide power production methods have both financial and ecological implications, especially in light of the demand for sources of alternate energy. Energy sources such as coal, oil, LPG, hardwood, everything of the fossil fuels (including biogas and hydro), wind, wave power, every form of energy used for energy supply (including limited), nuclear energy, energy production, and the reuse of energy are the types of utilities that need the most interest. Radiation continues to be released into the environment for hundreds of years after materials used as nuclear reactor fuel have been removed from their storage locations. How does the irradiation produced by those storage places affect the environment when it is unclear exactly how much or just how big of an impact it has [2].

Thermoelectric power plants are components that generate electrical current along the same path of operation when there is a temperature differential across their two surfaces. The following are a few investigations on the TEC (thermoelectric cooling) generator: i. A mechanism was created to directly convert electrical power into thermal energy during a study, and the power source was charged. A microcontroller monitored the maximum power point, and it reached 7.99 W of power. TE element 112708'dir TEC was used. The module in question has undergone heat and strength testing. ii. A TE generator for electrical power was installed on the stove's top as part of a study. The electricity was obtained using a hot burner as the source and a cold source (the ambient temperature of the air inside the chamber). iii. A device for converting geothermal power into electricity and running a microcontroller-controlled thermoelectric power plant had been evaluated as part of an investigation into sources of clean energy. iv. In an investigation of thermoelectric power plants, power is produced by focusing geothermal and solar power onto microcontroller controls as electrical power. Thermoelectric TEC-Peltier modules were created as part of this research by converting electrical power from the thermoelectric generator's application into heat energy. When modules are coupled in series, extremely low heat exchanger yields have been attained within the necessary voltages to be used at the place of installation. Geothermal power can be transformed into electrical power, which is widely available and simple to obtain by boosting the number of components utilised in the process [3].

2. Literature review

Thermal fluid conditions, input power, and the number of modules (with the same total input power) are changed to achieve the research's goal. The cooler is then optimised from the viewpoint of COP (coefficient of performance), exergetic efficiency, and cost per unit of cooling. In [4], they tested a solar thermoelectric adsorption refrigeration device for eight days as part of an experiment. During the day, the Peltier effect and a thermoelectric device made the room cooler. On the other hand, an adsorption cooling device is what makes the room cool at night. The total COP of the system was the sum of the COPs of the two combinations, which were thermoelectric cooling (COP = 0.152) and adsorption (COP = 0.131). [5] conducted an experiment on thermoelectric cooling devices for a spatial treatment application inside a building. They carried out the laboratory testing in such a way that it was as close as possible to the optimum real state of the building's environment. A thermoelectric cooling gadget was put together in the lab for the test work. They found that eight units of the ultra-thermoelectric cooling device could produce a cooling effect of 220 W with a COP of 0.48 and an input current of 4.9 A on each module.

In [6], the performance of a thermoelectric air-cooling system was compared to that of traditional vapour compression and vapour absorption air conditioning. It was found that the thermoelectric system's coefficient of performance (COP) was lower than that of the traditional system. In [7], we compared the performance behaviour of three domestic refrigerators based on vapour compression, vapour absorption, and thermoelectric systems at similar capacities. Based on the experimental data, Bansal and Martin reported that the COP of the VC system was relatively higher (COP = 2.57), followed by the thermoelectric system (COP = 0.68), and the absorption system (COP = 0.46). The system based on it has a relatively low COP, but the technology is quite advanced and reliable, and the price of the cooling system is quite favourable. A series of prototype thermoelectric coolers were designed and examined, utilising various thermoelectric modules. The performance and effectiveness of these coolers were then compared, taking into consideration their coefficient of performance (COP), temperature stability, heat pumping capacity, and cooling rate [8].

Based on the rules of non-equilibrium thermodynamics [9], the study focused on cyclic models of one-stage and two-stage semiconductor thermoelectric cooling with one and two stages. In [10], a two-stage thermoelectric refrigerator with two structural configurations was investigated. The analysis focused on evaluating the performance of a thermoelectric refrigeration system with multiple elements within a limited time frame, based on the principles of thermodynamics. In order to enhance and optimise the cooling load and coefficient of performance (COP), an optimisation process was conducted to determine the ideal ratio between the heat transfer surface area of the high-temperature side and the total heat transfer surface area of the heat exchangers [11].

3. Implementation of proposed research

The MPPT controller for charging will receive electricity from the photovoltaic cell. By fast charging the power source, the MPPT controller for charging makes sure that the electrical loads receive the greatest amount of current that may be utilised. Maximum power point (MPP) may be thought of as the optimal voltage where the loads get the most power while experiencing the fewest losses. Peak power voltage is another name for this. Since the charging controller will monitor the highest peak power, it will operate more effectively. The respective current and voltage sensors will sense the current and voltage. A microcontroller will help display the power, current, and voltage through an LCD display. The Buck-Boost converter will change the power ratings as per the battery ratings.

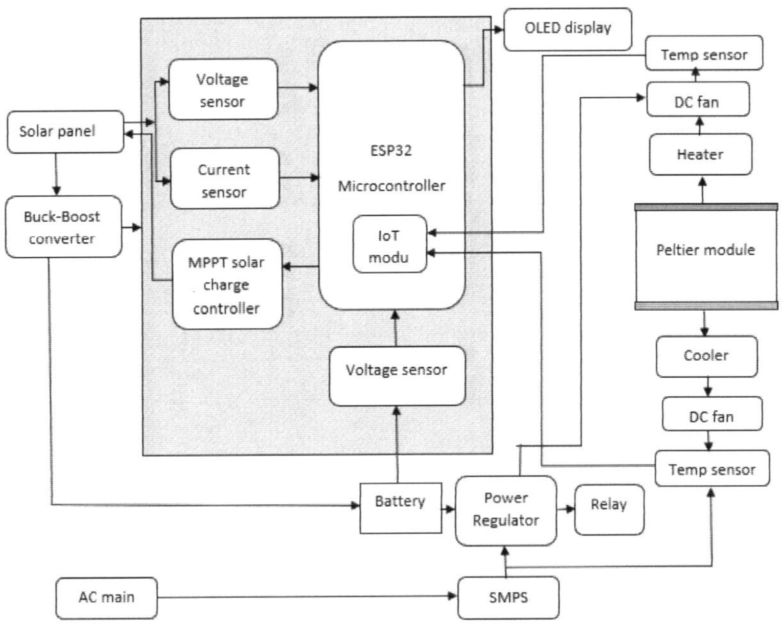

Figure 1. Schematic block diagram of an innovative experimental analysis of optimised solar-powered Peltier-based air cooler and warmer in medical systems.

The output from the battery is given to the PWM controller, which alters the current flow and helps to change the temperature of the Peltier module. The effect known as Peltier is most commonly used in industry through Peltier elements. These are linked in succession. The electrical conductivity of both substances differs as a result of being contaminated with foreign atoms. A portion of the electrical power is turned into heat when a voltage is supplied to the semiconductor's terminals. As energy is removed from the Peltier component by running an electrical current across it, the front of the device gets cold, while the opposite side emits heat into the surrounding space. Hence, one side is cooled, which can be used as a medical refrigerator, and the opposite side simultaneously heats up, which can be used for food warming.

The temperature sensors are connected to each side of the Peltier module to measure the respective temperatures of the cold and heat sides. The whole system is connected to an IoT module, where we can access the real-time controlling and monitoring of the whole system as shown in figure 1.

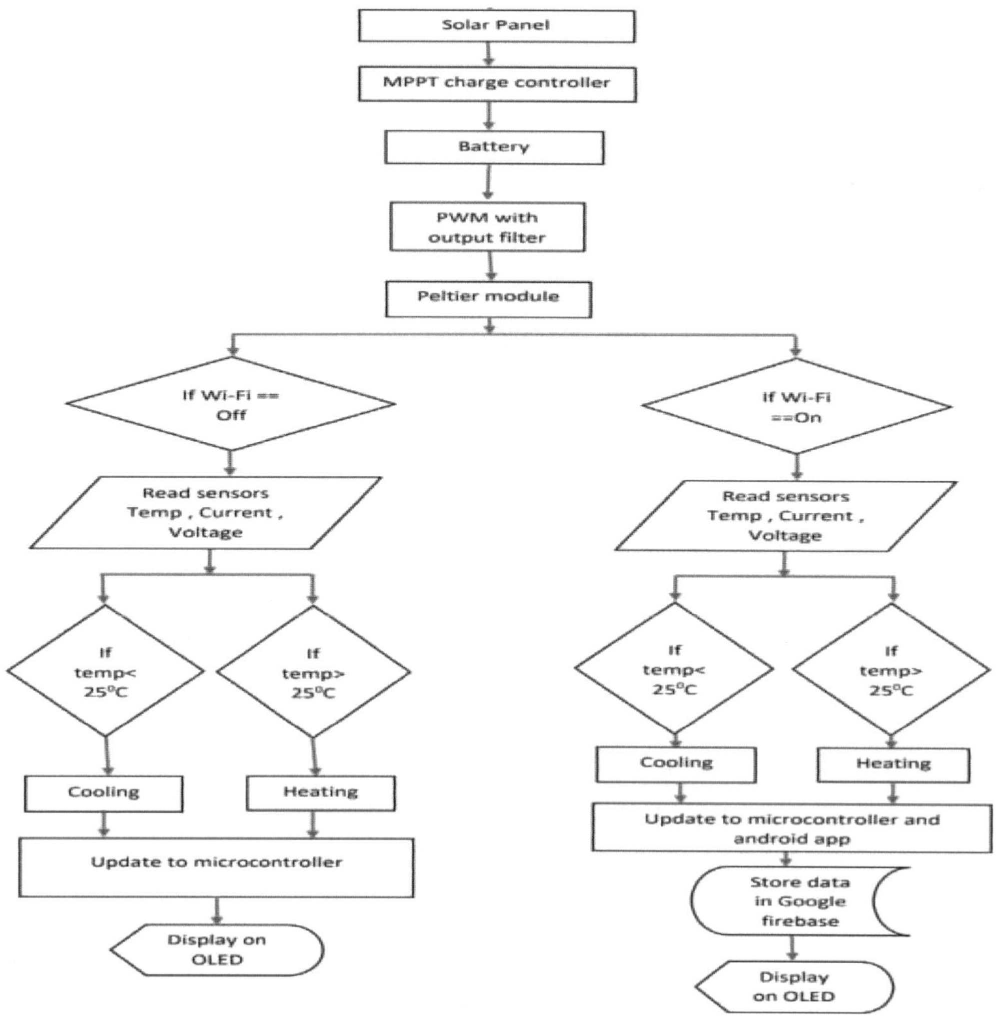

Figure 2. Flowchart of an innovative experimental analysis of optimised solar-powered Peltier-based air cooler and warmer in medical systems.

As explained in the block diagram, the flow of the model is explained above. The power is supplied to the microcontroller and battery with the help of Solar panels. Solar panels will charge the battery through the MPPT charge controller circuit. The MPPT charge controller uses algorithms such as the perturb and observe methods to extract the maximum peak power from the solar panels. It uses if-then conditions and feedback to extract maximum power from solar panels. Batteries will be basically rechargeable, such as lithium-ion or lead-acid batteries. Since the power from the battery is constant, it is needed to supply variable current to the Peltier to change its temperature. So, it needs a PWM output filter. Then the power is supplied to the Peltier through a PWM filter. The system works on two conditions, i.e., whether the Wi-Fi is connected or not

In the first condition, where Wi-Fi is not connected, the whole system is offline or manually controlled. All the sensors, i.e., the voltage sensor, current sensor, and temperature sensor, will read the respective values, which will be sent to the microcontroller. The microcontroller checks whether the temperature is less than room temperature (250 °C) or more than room temperature. If it is higher than the room temperature, it results in heating, and if it is lower, it results in cooling. All these readings are updated by the microcontroller, and the microcontroller sends the signals to the OLED display, through which the output of all three sensor values is displayed.

In the second condition, where Wi-Fi is connected, the whole system is online or IOT controlled through an Android app. All the sensors, i.e., the voltage sensor, current sensor, and temperature sensor, will read the respective values, which will be sent to the microcontroller. The same is updated for the app. Here, Google Firebase stores the data for each state. The microcontroller checks whether the temperature is less than room temperature (250 °C) or more than room temperature. If it is higher than the room temperature, it results in heating, and if it is lower, it results in cooling. All these readings are updated on the Microcontroller mobile app and Google Firebase. The microcontroller sends the signals to an OLED display, through which the output of all three sensor values is displayed, as shown in figure 2 and figure 3 shows the circuit diagram of the proposed work.

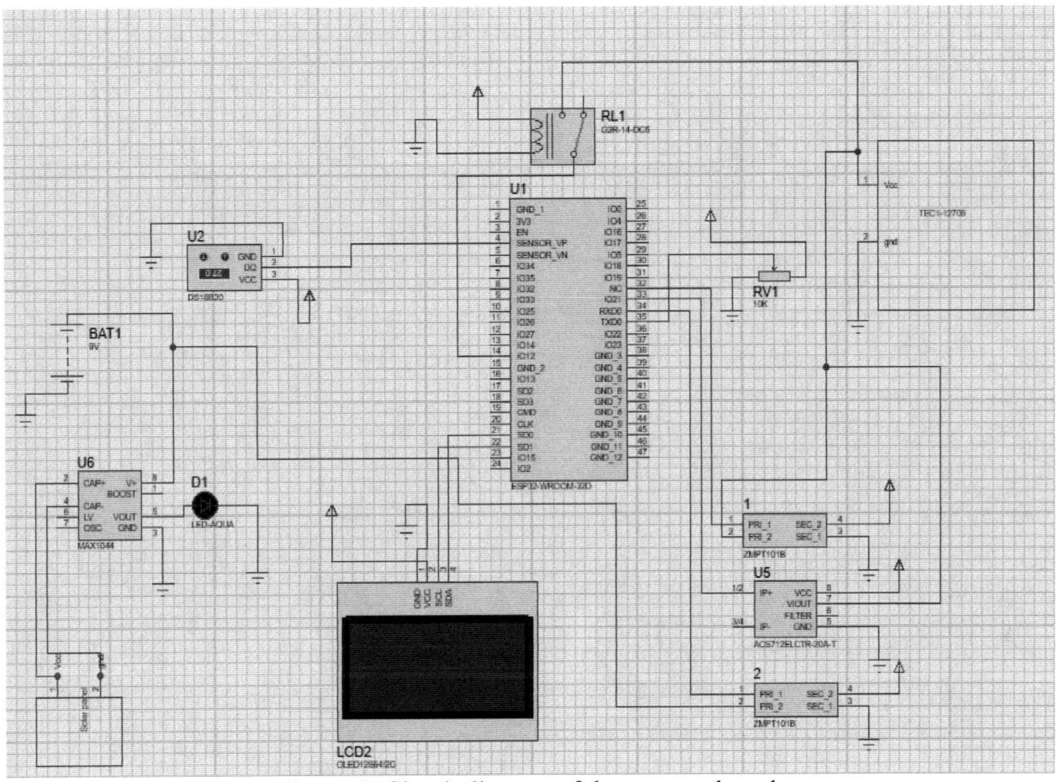

Figure 3. Circuit diagram of the proposed work.

4. Results and discussion

Once the whole hardware component is installed with proper connections, the system can be operated and the desired effect can be obtained. The variation in temperature on the heating and cooling sides of the Peltier module is recorded with respect to time and temperature, and a graph is plotted.

4.1. Temperature at Heating Side

The temperature versus time characteristics on the heating side are obtained as in Fig. 4. As we can see in Table 1, the maximum temperature on the heating side is 350. Which is the maximum limit set based on the requirements of the warmer. Whereas the temperature on the heat side can go up to 640 Celsius if required. The temperature versus time characteristics on the heating side are obtained as in Fig. 4. As we can see in table 1, the maximum temperature on the heating side is 350. Which is the maximum limit set based on the requirements of the warmer. Whereas the temperature on the heat side can go up to 64°C if required

Table 1. Temperature vs time at heating side.

Air temperature at Heat side	
Temperature (t1) in °C	Time in (s)
25(Room temperature)	0
27.5	5
27.5	10
28	15
30	20
31.5	25
31.5	30
33.0	35
34.0	40
35.0 (Max ref temp)	45

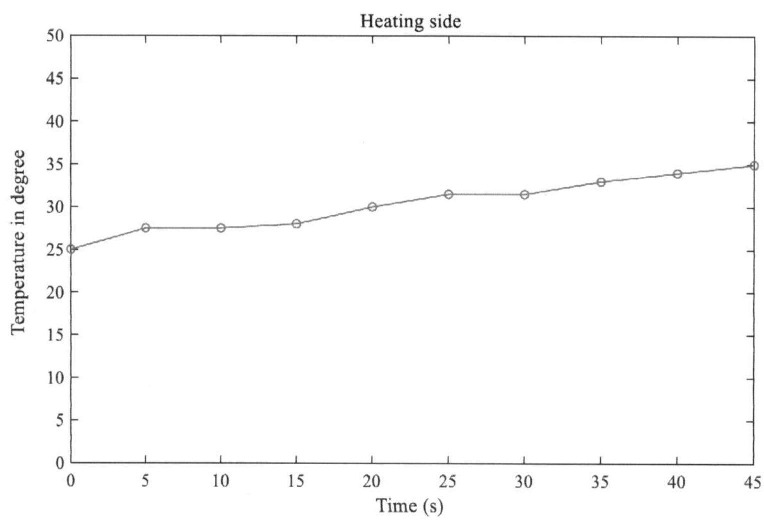

Figure 4. Graph of temperature v/s time at the heating side.

The model can achieve the maximum temperature in less than 1 minute, which is too quick as compared to the conventional forms of heaters and warmers available in the market. This not only allows us a quick heating capacity but also requires less power as compared to other conventional heaters.

4.2. Temperature at Cooling Side
The Temperature versus Time characteristics on the Cooling side are obtained as in Fig. 5. As we can see in table 2, The minimum temperature on the Cooling side is 150. Which Is the minimum limit set based on the requirements of the warmer? Whereas the temperature on the cool side can go up to 110°C if required.
The model can achieve the minimum temperature in less than 1 minute, which is too quick as compared to the conventional forms of Coolers available in the market. This not only allows for quick cooling but also requires less power as compared to other conventional Coolers.

Table 2. Temperature vs time at cooling side.

Air temperature at cooling side	
Temperature (t2) in °C	Time in (s)
25 (Room temp)	0
24	5
23	10
23.5	15
21	20
19.5	25
17	30
16.5	35
15	40
13(Max cooling)	45

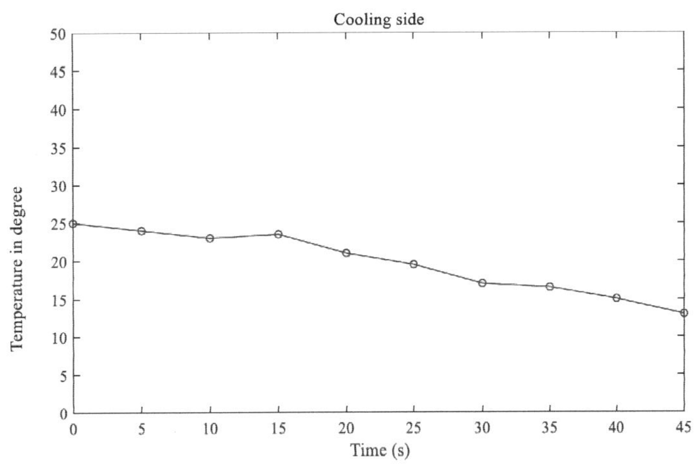

Figure 5. Graph of temperature v/s time at the cooling side.

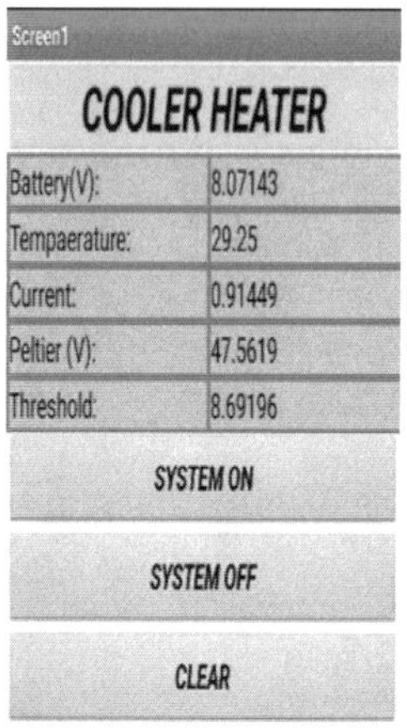

Figure 6. Final working model and android app.

5. Conclusion

The temperature attributes of both the heat-radiating and heat-absorbing sides of the Peltier module were determined based on the readings from the model. The voltage versus time graph exhibited a consistent outcome throughout the duration of the prototype's performance evaluation. The experiment was carried out by manipulating the concentration of solar energy over a period of time. Based on the findings obtained from the conducted experiments, it was observed that the upper limit of temperature attainable within the hot chambers is approximately 35°C. Conversely, the lower limit of temperature attainable within the alternate chambers is approximately 13°C, thus representing the colder of the two. Therefore, the heat energy extracted by the Peltier module is effectively utilised for heating purposes, thereby serving the dual function of heating and cooling within one module.

6. References

[1] Arjun K G B, Pruthviraj B G, Chethan K Y K and Rashmi P 2017 *RTEICT 2017 - 2nd IEEE International Conference on Recent Trends in Electronics, Information and Communication Technology, Proceedings* p 17504195

[2] Attavane P, Arjun G B, Radhakrishna R and Jadav S R 2017 *RTEICT 2017 - 2nd IEEE International Conference on Recent Trends in Electronics, Information and Communication Technology, Proceedings* p 17504123

[3] Kumar Gb A, Sushma S, Priyanka L, Vijay S G and Thouqhir Pasha G A 2018 *2018 3rd IEEE International Conference on Recent Trends in Electronics, Information and Communication Technology, RTEICT 2018 – Proceedings* p 19410680

[4] Wen S, Zhang G, Dan Y, Wang D and Deng M 2012 *2012 International Conference on Advanced Mechatronic Systems, ICAMechS* p 452-457

[5] Du J, Yang F and Wu J 2012 *IEEE International Symposium on Industrial Electronics* p 12853396

[6] Chen S, Zuo J and Xie D 2010 *Proceedings of the World Congress on Intelligent Control and Automation (WCICA)* p 11512131

[7] Chavan S S 2022 *Int J Res Appl Sci Eng Technol* **10** 1043

[8] Twaha S, Zhu J, Yan Y and Li B 2016 *Renewable and Sustainable Energy Reviews* **65** 698

[9] Alaoui C and Salameh Z M 2001 *LESCOPE 2001 - 2001 Large Engineering Systems Conference on Power Engineering: Powering Beyond 2001, Conference Proceedings* p 7042360

[10] Zaferani S H, Sams M W, Ghomashchi R and Chen Z G 2021 *Nano Energy* **90** 106572

[11] Tian X X, Asaadi S, Moria H, Kaood A, Pourhedayat S and Jermsittiparsert K 2020 *Energy* **208** 118428

Design of 3-Level T-type NPC Inverter for EV Applications

Balamurugan M[1*], Arjun Kumar G B[1], Sujit Kumar[1], and Sridhar G[1]

[1] Department of Electrical and Electronics Engineering, Dayananda Sagar College of Engineering, Bengaluru, Karnataka, India

balamurugan-eee@dayanandasagar.edu

Abstract. In recent years, there has been a significant demand for lightweight converters that provide exceptional control performance while generating minimal acoustic noise. This has resulted in higher switching signals for hard- switched 2-level low voltage 3-phase inverters. However, the increased switching frequencies have a negative impact on the inverters' efficiency, requiring the use of expensive switch technology to maintain high efficiency. This paper aims to tackle these challenges by introducing the design and implementation of a extremely effective 3-Level T-Type Neutral Point Clamped (NPC) inverter designed for Electric Vehicle (EV) applications. 3-Level T-Type NPC inverter associates the strengths of both NPC and T-Type inverters. Similar towards the NPC, it offers a lower breakdown voltage for the switching devices, thereby reducing switching losses. Additionally, it shares the characteristics of the T-Type inverter, resulting in fewer conduction losses. To validate the feasibility of the converter, simulation is performed using MATLAB software, followed by the development and testing of a lab-scale prototype. The proposed three-level T-type NPC inverter have several merits over traditional 2-level inverters, including reduced distortion, switching losses, and improved efficiency. These benefits contribute to enhanced motor drive performance, reduced electromagnetic interference (EMI), and extended battery life in EVs.

1. Introduction

The efficiency and performance of a two-level inverter are subject to limitations due to the fixed output voltage determined by ±VDC/2. Additionally, the chosen switching method affects the inverter's efficiency and performance. Implementing the six-step modulation technique reduces switching loss but leads to increased copper loss and torque ripple because presence of low-frequency harmonics [1].

To address the harmonic issues of the six-step method, the widely used solution is pulse width modulation (PWM) in many applications. The copper loss and torque ripple are reduced by employing PWM compared to the six-step method, thanks to the higher switching frequency. However, this approach leads to increased switching loss [2].

As the switching frequency and DC link voltage increase, the influence of switching loss becomes more significant. To overcome the limitations in efficiency and performance associated with a two-level inverter, it is advisable to adopt a 3-level T-type NPC inverter. Selection provides additional improvements in energy efficiency, reliability, power output, and density.

A 3-level NPC inverter belongs towards the category of multilevel inverters designed to convert DC power into AC power. The term "three-level" signifies the utilization of three levels of DC voltage (positive, zero, and negative) to generate a stepped AC voltage waveform. The T-type configuration refers to how the DC voltage sources are interconnected within the inverter [3,4].

IOP Conf. Series: Materials Science and Engineering 1295 (2023) 012002 doi:10.1088/1757-899X/1295/1/012002

When designing a 3-level T-type NPC inverter for electric vehicle (EV) applications, several factors need to be taken into account. These factors encompass current and voltage specifications of the inverter, power demands of electric motor, size and weight limitations of vehicle, and the desired efficiency and reliability of the inverter.

2. Circuit description

The simulation process involves the utilization of MATLAB software, a robust and versatile tool for simulating inverters, including the sophisticated two-level Neutral Point Clamped (NPC) inverters. This software offers a comprehensive array of features and functions that facilitate the analysis and optimization of inverter performance. Employing MATLAB for inverter simulation provides numerous advantages, such as a broad spectrum of functions capable of accurately modelling various types of inverters, notably the two-level NPC inverters. The simulations conducted using MATLAB are prominent for their high precision, enabling accurate predictions of inverter performance across diverse operating conditions. Furthermore, the versatility of MATLAB extends to its applicability in simulating inverters for a wide range of applications, renewable energy systems, electric vehicles, and more [5,6].
The control strategy utilized to regulate output voltage and frequency of inverter is sinusoidal Pulse Width Modulation (PWM) technique. This method involves employing either a microcontroller or a dedicated SPWM generator circuit to produce SPWM waveforms. These waveforms are then applied to the inverter switches, specifically MOSFETs. Through rapid switching the switches on and off at a high frequency, synchronized with the SPWM signals, the inverter generates an output voltage waveform that closely resembles a sinusoidal waveform [7].
In order to serve as the control interface for 3-level T- type NPC inverter, the Arduino Uno board is employed. To facilitate this functionality, the Arduino Uno must be programmed to generate the requisite control signals essential for the proper operation of the inverter.

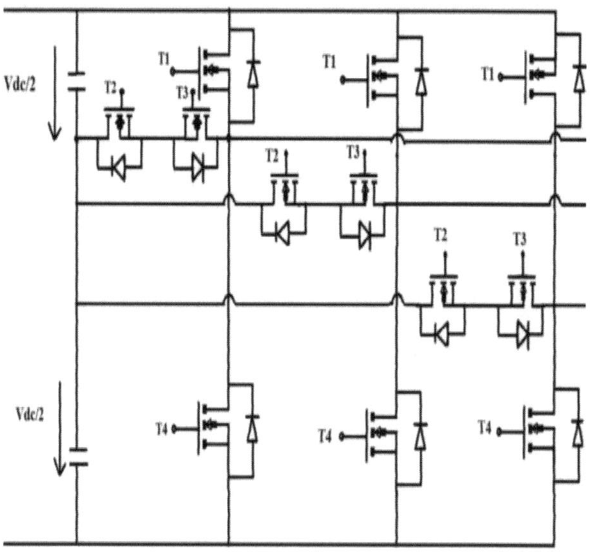

Figure 1. A schematic representation of T-type inverter.

Figure.1 can be used to depict the 3-level T-type inverter. This topology extends the traditional 2-level VSC configuration by integrating an dynamic bidirectional switch that connects to midpoint of dc-link. T-type VSC configuration resembles a rotated character "T" in one of its legs, giving it its distinctive name. By employing this T-type structure, a more advanced and improved system can be realized.
The verified effectiveness of the 3LTNPCI is supported by precise measurements carried out on a hardware prototype. This prototype demonstrates an impressive efficiency of 99% and a power capacity of 10 kilowatts. Significantly, this cutting-edge prototype showcases an exceptional capability to adapt its switching signals effortlessly transition among 2-level and 3-level modulation modes while in

process. Consequently, it allows for a direct assessment of the effects on passive components, including the load machine and EMI filter, facilitating a thorough evaluation of their impact on losses.
Owing to its remarkable properties of minimal conduction losses and negligible switching losses, the 3LTNPCI demonstrates exceptional efficiency, particularly when deployed within the medium switching frequency range of 8-24 kHz, commonly utilized in various industrial applications. Importantly, this impressive performance is accomplished without resorting to the utilization of costly Sic (Silicon Carbide) devices, which may be required if chip dimensions are selected to achieve equivalent conduction losses. Thus, 3LTNPCI emerges as a highly viable alternate to 2-level converters in low-voltage scenarios, where the attainment of elevated switching frequencies and utmost efficiency constitutes a pivotal objective [8,9].

3. T-type topology implementation

Circuit diagram presented in Figure 1 illustrates the essential layout of the 3-Level T-type NPC Inverter (3LTNPCI). Unique configuration expands upon the conventional 2-level VSC by incorporating a dynamic bidirectional switch at the core of the DC link. In low-voltage applications with a Vdc of 650 V, it is common practice to use up-side and down-side switches (T1/D1 and T4/D4) constructed with 1200-V MOSFETs/diodes to fully block the DC-link voltage. Nevertheless, its important near note that the bidirectional switch located at centre point of DC link is specifically designed towards handle and block only a portion of DC-link voltage. Therefore, its possible to utilize devices with a less voltage rating. In this specific scenario, two 600-V MOSFETs, each with antiparallel diodes, are employed (as shown in figure 2). By operating at a lesser stalling voltage, the centre switch experiences minimal switching losses and adequate conduction losses, despite having 2 devices coupled in series [10].
Unlike 3-level Neutral Point Clamped (NPC), the T-type configuration stands out by eliminating the need for a series association of devices to completely stall DC-link voltage, Vdc. In NPC topology, the transition between positive (P) and negative (N) DC-link voltage levels is often avoided during switching to prevent uneven voltage distribution when both series-connected MOSFETs turn off simultaneously. However, such undesired effects are not present in the T-type topology. As a result, there is no need for additional measures to prevent these transitions or confirm transient voltage matching amongst the series-connected MOSFETs [11].

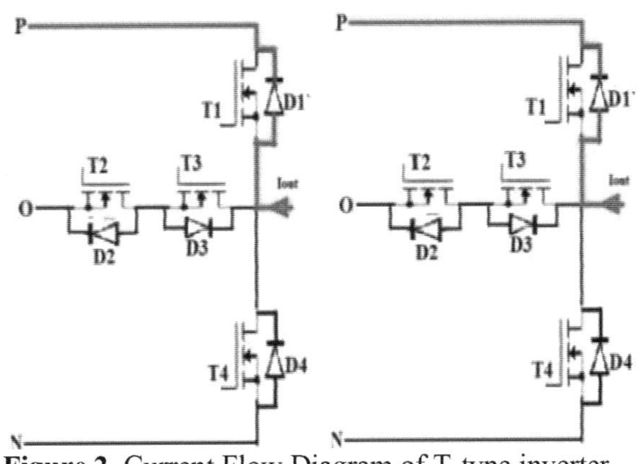

Figure 2. Current Flow Diagram of T-type inverter.

The utilization of a bidirectional switch to achieve a transition toward the midpoint position can be effectively executed through the utilization of MOSFETs. This implementation may be achieved by configuring the MOSFETs in either a common emitter arrangement (referred to as a) or in a common collector configuration (referred to as b). A further benefit of utilizing one 1200-V devices to stall the entire DC-link voltage is the reduction in conduction losses, especially when considering bipolar devices. In the 3LTNPCI configuration, when output is coupled to either the positive (P) or negative (N) side, only one device incurs a forward voltage drop, unlike NPC topology where 2 devices are continuously coupled in series. This significantly decreases conduction losses, producing 3LTNPCI configuration an appealing option at lower switching frequencies [11].

Traditionally, the 3-level NPC topology was primarily designed for medium voltage applications. In those scenarios, connecting in series was essential because of the limitations of available devices in terms of voltage-blocking capacity. Although the series connection of bipolar devices led to higher conduction losses, it was justified towards improved voltage and power handling capabilities. However, context of low-voltage applications, this advantage becomes inconsequential. Modern devices offer appropriate voltage ratings and higher switching speed, eliminating the necessity for series connections and thereby mitigating the associated conduction losses. Enhancing the phase currents is essential for achieving higher output power. However, it is important to minimize conduction losses, as they can restrict the power range. In this regard, the T-type topology emerges as a superior option for low-voltage applications that employ bipolar devices. The detrimental impact of conduction losses can be effectively mitigated by selecting the T-type configuration, enabling optimal power performance.

T-type topology is not limited towards low-voltage applications but is also employed in medium-voltage scenarios, called as the Neutral Point Piloted (NPP) converter. In medium-voltage applications, a single device's voltage-blocking capability is insufficient to stall the entire dc-link voltage. To address this limitation, a series connection of two MOSFETs is utilized for each of the switches T1 and T4. Additionally, specialized active gate drive units are required to ensure voltage balancing during both transient and steady-state conditions. However, this configuration increases implementation complexity and leads to higher conduction losses because of series association of bipolar devices. Consequently, it is important to note that the T-type topology is generally not suggested for less voltage applications.

3.1. Switch configuration

The arrangement of a bidirectional switch using a pair of 600-V MOSFETs can be accomplished through two main configurations: common collector and emitter configuration.

Common emitter configuration (figure 2(a)) requires an added remote gate drive supply voltage for each phase of the bridge, resulting in a total of three extra gate drive supplies equated to 2-level VSC. However, by utilizing the common collector configuration (figure 2), number of gate drive provisions may be reduced. In this configuration, the second MOSFET, T2, shares a common emitter with up-side switch, T1, and must be powered by the remote gate driver voltage of T1. Furthermore, emitter of next 600-V MOSFET is coupled to center point voltage level. In a three-phase topology, all three MOSFETs, T3, T4, and T5, stake a common emitter, requiring only remote gate driver source is shown in Table 1. Gate charges of 600-V and 1200-V MOSFETs are almost identical, the power rating of remote gate driver source for up-side switch, T1, does not need to increase. Commutation and modulation strategy employed ensures that T1 and T2 are never simultaneously switched in identical modulation cycle, which can be elaborated in following segment.

However, it is important to include six extra gate driver integrated circuits (ICs) and 6 added digital isolators for switch signals, it may result in a slight increase in costs. Nevertheless, equated to 3-level Neutral Point Clamped (NPC) topology, elimination of clamping diodes decreases required number of diodes after 18 to 12. Additionally, decrease in number of extra remote gate driver signals from 6 to 1 signifies the enhancement and has potential to reduce prices.

Table 1. Switching state of t-type NPC inverter.

State	Vout	T1	T2	T3	T4
1. P	+Vdc/2	ON	ON	OFF	OFF
2. 0	0	OFF	ON	ON	OFF
3. N	-Vdc/2	OFF	OFF	ON	ON

IOP Conf. Series: Materials Science and Engineering 1295 (2023) 012002 doi:10.1088/1757-899X/1295/1/012002

3.2. Commutation

The Process of switch commutation in 3LTNPCC (Three-Level T-type NPC Converter) configuration necessitates careful consideration. This configuration allows for output from a bridge phase to be coupled to either the positive (P), neutral (0), or negative (N) voltage level of DC link, has illustrated in Figure 3. To attain positive voltage level, switch T1 can be closed. For neutral level, switches T2 & T3 can be closed. Similarly, for the negative level, switch T4 can be closed, irrespective of the current direction. However, this approach relies on a commutation sequence that is contingent upon the direction of the current. Fortunately, there exists a simpler and more independent commutation strategy. To attain the chosen output voltage level, simplified commutation strategy can be employed in the 3LTNPCC configuration. By closing specific combinations of switches, such as T1 and T2 for positive voltage level, T2 & T3 for neutral level, and T3 & T4 for negative voltage level, current naturally flows through appropriate division regardless of its direction. Table I provides a reference for the specific switches that need to be bolted to attain chosen output voltage. To avoid short circuits in the DC link, a simple delay in turning on all switches is sufficient. By utilizing these switch signals, the modulation used in the 3LTNPCC configuration closely resembles the modulation employed in the three-level NPC topology (P→0) for (a) positive & (b) negative output current, current commutation through switching conversion (0→P) for (c) positive & (d) negative output current.

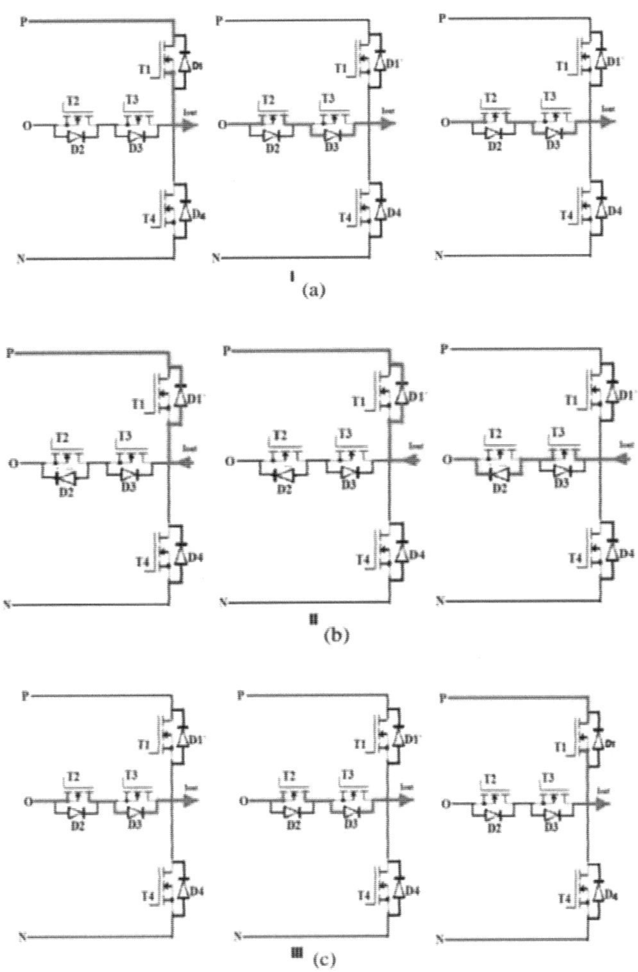

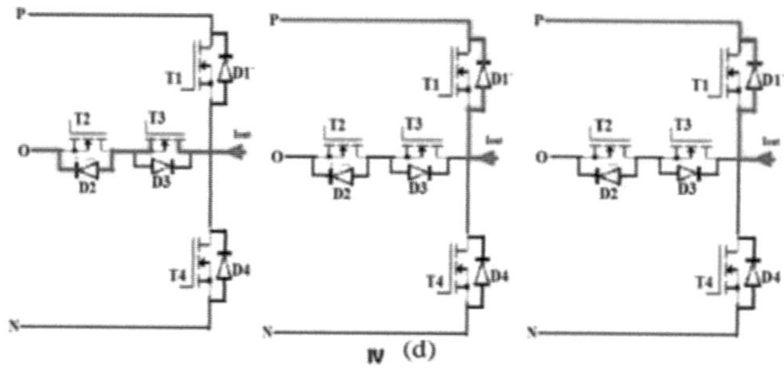

Figure 3. current commutation through switch transition.

To demonstrate switching transitions in 3LTNPCC configuration, can refer to figure 3(a). In this scenario, when output phase is coupled to positive voltage level with switches T1 & T2 closed, a positive output current requires a commutation to neutral level (0). This transition can be achieved by opening switch T1. Following a turn-on delay, switch T3 is closed. As switch T1 turns off, the current naturally flows through T2 and diode D3, reaching the neutral level. In order to transition from the neutral level (0) back to the positive level (P) in 3LTNPCC configuration, first step is to open switch T3. Following a turn-on delay, switch T1 remains then closed. Through turn-off T3, in case of a positive phase current is illustrated in figure 3(c), the current continues to flow through diode D3 & commutates to positive voltage level upon turn-on T1. Similarly, for negative phase current is shown in figure 3(d), the current commutates to diode D1 through the turn-off of T3. This principle applies to all the enduring switching conversions.

3.3. Control strategy

To address the issue of higher harmonics and lower efficiency in the inverter outputs, SPWM technique is utilized. This technique allows generation of a sinusoidal output waveform by continuously switching a three-level T-Type NPC inverter. The SPWM process involves comparing a triangular carrier waveform with the desired sinusoidal waveform within the inverter circuitry. In the three-phase configuration, three sinusoidal signals with phase angles of 0, 120, and 240 degrees are generated. These sinusoidal signals are then compared with positive carrier and negative carrier waveforms, resulting in the generation of positive and negative PWM pulses. These pulses are subsequently fed to the switching devices in the inverter, which is organized into three separate legs. Each leg requires a specific set of high and low pulses. To monitor switching signals across all three phases of switching devices, a display is utilized. Arrangement of the switching signals is designed in a mode that ensures alternating switching between positive and negative signals occurs at regular intervals for each switch.

3.4. Output signals

The present state and the electric potential and current wave patterns emanating from the inverter are closely observed through the utilization of the MATLAB/Simulink scope.

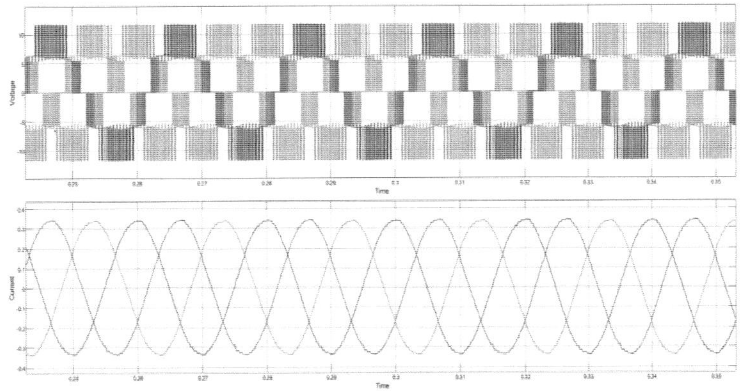

Figure 4. Voltage and current waveform of inverter.

Resulting waveforms of line voltage and current are visually showed in the figure 4. Subsequently, the produced alternating current is fed into a resistive load, and the operational characteristics of said load are continuously monitored to discern any alterations in the input parameters. The waveforms exhibited above are the outcomes produced by the MATLAB/SIMULINK software during the process of designing the three-phase SPWM inverter, with each distinct waveform being generated at different stages of modulation is shown in figure 4

3.5. Hardware prototype

In summary, the Arduino UNO amplifies the pulses from Simulink and generates PWM signals. These signals are then passed through the gate driver circuit using optocouplers. The optocouplers provide electrical isolation and control the MOSFETs in the inverter circuit. The MOSFET switches, controlled by the PWM signals, produce the desired three-level output waveform. The entire circuit is powered by a 12V DC supply is displayed in figure 5

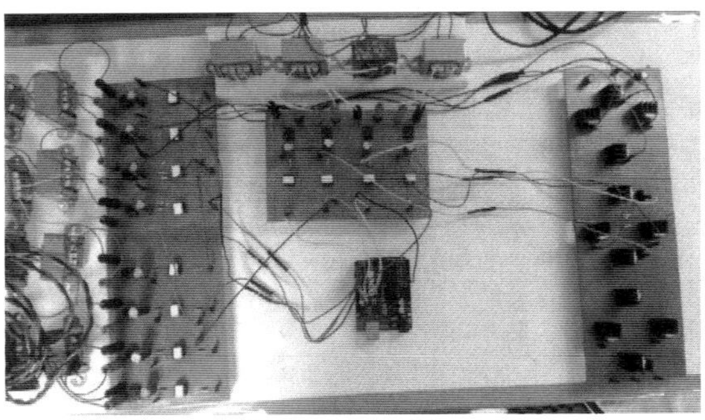

Figure 5. T-type inverter prototype.

3.6. Hardware prototype

In summary, a 3-level T-type NPC inverter with 12V DC input generates an output voltage waveform with three distinct voltage levels (-Vdc, 0, and +Vdc) through pulse width modulation. This waveform shown in figure 6 exhibits reduced harmonic distortion, smoother voltage profile, and enhanced voltage control, contributing to improved efficiency and performance in relevant applications.

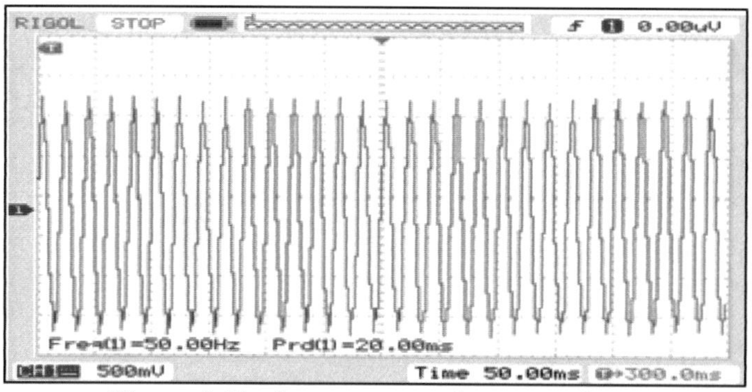

Figure 6. Hardware output voltage waveform.

4. Conclusion

This research paper introduces the Three-Level Three- Phase NPC Inverter (3LTNPCI) as an efficient solution specifically designed for low-voltage applications, providing an alternate to conventional Two-Level Voltage Source Converter in medium switching frequency scenarios. Within the frequency range of 4 to 30 kHz, the 3LTNPCI showcases exceptional performance. Its primary advantage lies in reduction of switching losses by halving the commutation voltage equated to the 2-level topology. While conduction losses remain relatively unchanged, the paper delves into the implementation of a bidirectional switch utilizing two anti-serial 415-V MOSFETs, which can be configured in whichever a common-emitter or common-collector arrangement. The paper emphasizes the advantages of incorporating additional isolated gate drive supplies in the common-collector configuration. It also presents a simplified commutation scheme that operates independently of the current direction. To ensure an accurate assessment of the losses in the 3LTNPCI, the paper introduces an algorithm that takes into account the loss optimum clamping scheme and temperature-dependent characteristics of loss components. Furthermore, research includes measurements conducted on a 1-kW prototype of the 3LTNPCI, providing empirical evidence that confirms the predicted efficiency of 99%.

5. References

[1] Vijeh M, Rezanejad M, Samadaei E and Bertilsson K 2019 *IEEE Trans Power Electron* **34** 9479
[2] Perez M A, Ceballos S, Konstantinou G, Pou J and Aguilera R P 2021 *IEEE Open Journal of the Industrial Electronics Society* **2** 224
[3] Hussein A S, Jayan V and Ghias A 2019 *Proceedings of the IEEE International Conference on Industrial Technology* p 18797937
[4] Hussein A S and Ghias A 2020 *IECON Proceedings (Industrial Electronics Conference)* p 20193823
[5] Gundebommu S L 2018 *2018 International CET Conference on Control, Communication, and Computing, IC4* 218 p 18249602
[6] Lakshmi G S, Rubanenko O, Swarupa M L and Deepika K 2020 *Proceedings - 2020 IEEE India Council International Subsections Conference, INDISCON* p 20426016
[7] Maheshwari R, Munk-Nielsen S and Busquets-Monge S 2013 *IEEE Transactions on Industrial Electronics* **60** 1861
[8] Chen H C, Tsai M J, Wang Y B and Cheng P T 2018 *IEEE Transactions on Industry Applications* **54** 2517
[9] Sedaghati F and Latifi S H 2018 *9th Annual International Power Electronics, Drive Systems, and Technologies Conference, PEDSTC* p 17746055
[10] Hasan N S, Rosmin N, Osman D A A and Musta'amal@Jamal A H 2017 *Renewable and Sustainable Energy Reviews* **80** 163
[11] Susheela N, Kumar P S and Sharma S K 2018 *IEEE Transactions on Industry Applications* **54** 2425

Autonomous Wheelchair for Indoor Navigation

Chandana H R[1], Deepak M[1], Megha Anil Padanad[1], and Satish B A[1*]

[1] Department of Electrical and Electronics Engineering, Dayananda Sagar College of Engineering, Bengaluru, Karnataka, India

satish.eee@dayanandasagar.edu

Abstract. Wheelchairs are a very beneficial tool for commuting used by physically challenged people. But not everyone can make the wheelchair move on their own and often depend on help for movement of the wheelchair, making them dependent on constant support for every movement. Even when the physically challenged individuals are indoors, they end up being the dependent. The idea being proposed in the paper is develop a wheelchair that can make the movement of the wheelchair either automatic with inbuilt indoor navigation for easy navigation along with voice-based commands and joystick control for easy control. The wheelchair proved to be extremely beneficial for the physically challenged, at a relatively low cost.

1. Introduction

Around 65 million people depend on wheelchairs for their daily mobility needs. There has been a growing recognition of the need to enhance the quality of life and independence for the physically challenged.

The conventional manual wheelchair, while a valuable mobility aid, poses limitations in terms of manoeuvrability and control, particularly within indoor environments.

Some of the major challenges that users find in using the wheelchair by themselves in an indoor environment are

- Navigating through tight spaces: The amount of free space for navigation in an indoor domestic setup is always limited making it difficult for the users to navigate by themselves
- Avoiding obstacles: Any domestic indoor setup would always have several obstacles along the path, each obstacle requiring the user to navigate around the obstacle. This continuous need of avoiding and navigating around obstacles can be tiring for the users.
- Accessing different areas of a building: At times some parts of the building could be very difficult to navigate by themselves as it might be too steep or have too many obstacles.

All the above-mentioned limitations not only impede the independence of wheelchair users but also contribute to feelings of frustration and dependency.

There have been many efforts to overcome these problems by motorized and smart wheel chairs designed in the past. In [1] authors discussed the development of a voice and gesture-based wheel chair, allowing the users the freedom to move the wheelchair just by the voice commands as well as the gestures. In [2] authors have designed a wheelchair which also provides control of the wheelchair just using their eyeballs. Such a control is very useful for users who have severe disability in movement of their head or neck to direct the wheelchairs. In [3] authors have developed a wheelchair that can be controlled just by the blinking patterns of the eye. In [4] authors used the eye movement to aid the wheelchair control. In [5] authors provided the control of a wheelchair using eye movement as well as voice commands. Several such similar automations have been developed and discussed in [6–9]. The

wheelchairs have also been automated towards avoiding obstacles and ease of navigation. In [10] authors used GPS based location for navigation of a wheelchair. But for indoor navigation GPS is not the most reliable method due to the short distances that the wheelchair needs to travel. In [11] authors used computer vision for object detection and navigation of the robots. All the above presented methods make the wheelchair a lot easier to use for the user. But when it comes to indoor navigation the movement of the wheelchair still needs continuous inputs from the user for its movement in all the above methods is not completely autonomous in an indoor environment.
The paper addresses this issue by proposing a completely autonomous wheelchair which can navigate from one room to another room within a house or a building just by the input of the destination. such a robot minimizes human intervention to minimum and hence makes it a lot more user friendly for the physically challenged.

2. Methodology

Indoor navigation has been studied extensively in various scenarios and methods. In [12] authors discussed the creation of indoor navigation for IoT based applications using Wi-Fi signals. In [13] authors used 802.11 signals along with access points to create an indoor map and use the same for navigation. In [14] authors presented the use of computer vision for indoor navigation and mapping. In [15] authors presented the use of geomagnetism and the information of crowdsourcing paths to develop indoor navigation paths. In [16] authors used infrared and radio waves for navigation. The above-mentioned techniques have proven to work effectively but the main drawback of them was found to be the requirement of extra hardware present inside the house to make the mapping effective, thus could lead to increase in the overall cost of the system. To make the proposed wheelchair as economical as possible the implementation of the indoor mapping was done using distance-based calculations. The distances between the rooms are always fixed and the knowledge of this distance was used to create a map for the system from one room to the other and the movement of the wheelchair was controlled based on this map.
The major components required for the implementation were

- Sensors for measuring distance: The distance travelled by the wheelchair was required to be able to follow the map during the indoor movement, thus a sensor was used to keep track of the distance travelled by the wheelchair.
- Motors: The movement of the wheelchair had to be motorized so that the human effort would be minimal, thus motors were fitted on the wheelchair to make it move from one place to another.
- Controller: A central controller to control the movement of the wheelchair based on the command received from the user and presence of obstacles and other parameters.
- Obstacle detection: The movement of the wheelchair had to be done such that it would not collide with any of the obstacles that might come in front of it. To enable this an obstacle detection sensor was installed in the wheelchair at multiple places, the data of which was sent to the controller to act upon as required.
- User interface by voice: To make the use of the wheelchair easy the wheelchair was designed to accept commands from the user using voice recognition. The input commands would be the names of the rooms that were pre-configured in the system making it easy for the user to move from one location to another just by a voice command.
- User interface by joystick: It might be needed for the user to do minor movements or adjustments in the position of the wheelchair which the voice activation might fail to achieve, for this as a fall-back alternative a joystick was also provided allowing the user to control and move the wheelchair easily.
- User interface using mobile: A mobile application was also provided to the user to help the caregivers to move the wheelchair in case the user is unable to control or move for any reason.

The multiple user interface options allowed the wheelchair to be very user friendly and easy to use both for the user as well as the caregiver. Figure 1 gives the block diagram of the entire system to describe the interconnections between the various components described.

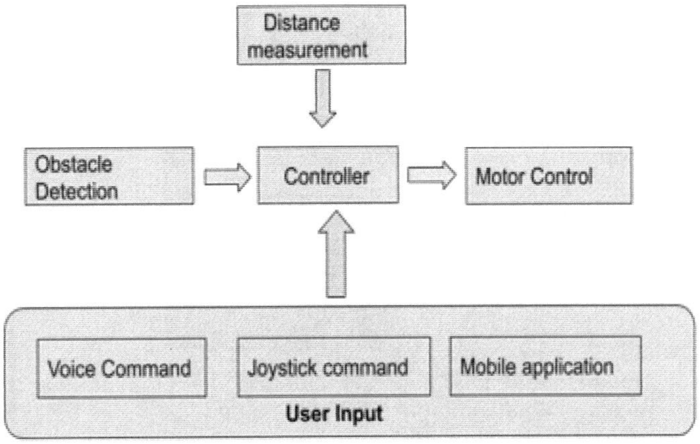

Figure 1. Block diagram.

3. Implementation

- Distance Measurement: To measure the distance travelled by the wheelchair IR sensors were fitted on the wheels of the wheel chair and the rotations of the wheel were used to calculate the distance travelled.
- Motors: Two DC motors were fitted on to the two rear wheels of the wheelchair. The rating of the motor was calculated to be able to carry a load of around 50 kg at a slow pace. Figure 2 show the motor used in the system.

Figure 2. Motor used.

- Controller: Arduino Uno was used as the controller in the prototype. This allowed the prototype cost to be lower and made it easy for testing the system under various conditions.
- Obstacle detection: Obstacle detection was done by placing ultrasonic sensors at multiple places on the wheelchair. The controller processed the inputs from all the sensors and acted accordingly.
- User Interface: The voice command was taken as input using a mobile itself as the input device. This reduced the hardware requirement in the wheelchair in turn reducing the cost of the system. The input from the user was relayed to the controller by Bluetooth interface between the controller and the user mobile. The joystick for control was provided on the handle of the wheelchair for ease of use. The mobile application was developed for android and it provided the user an option to choose as to which mode of control would he/she like to use on the wheelchair at a given moment. Figure 3 gives a sample image of the application.

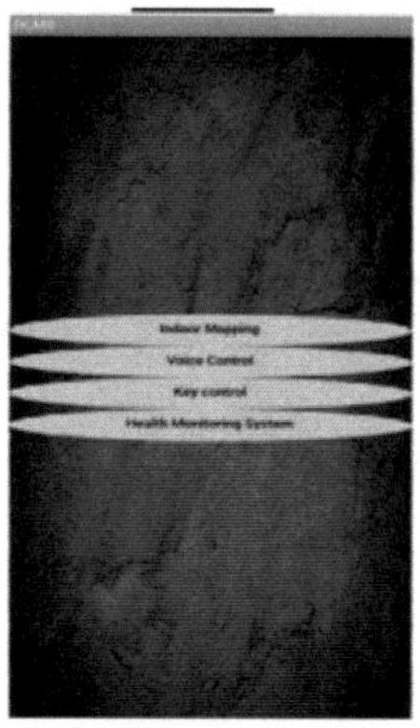

Figure 3. User interface in application.

Figure 4 gives a detailed flow chart of the usage of the wheelchair for the user taking input from a voice input device for destination for example a mobile phone and responding to the commands accordingly.

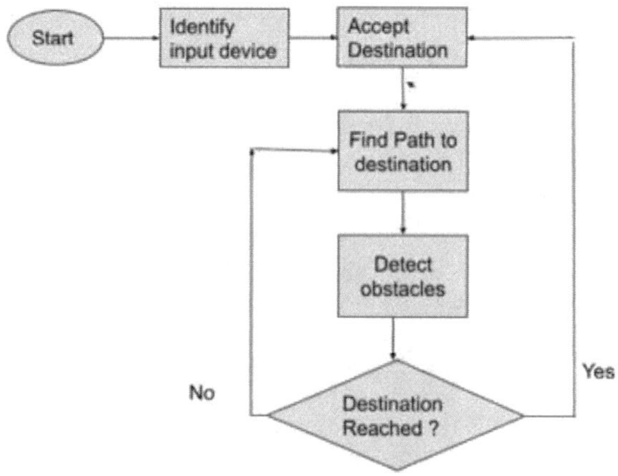

Figure 4. User interface in application.

Figure 5 is the final prototype developed as described.

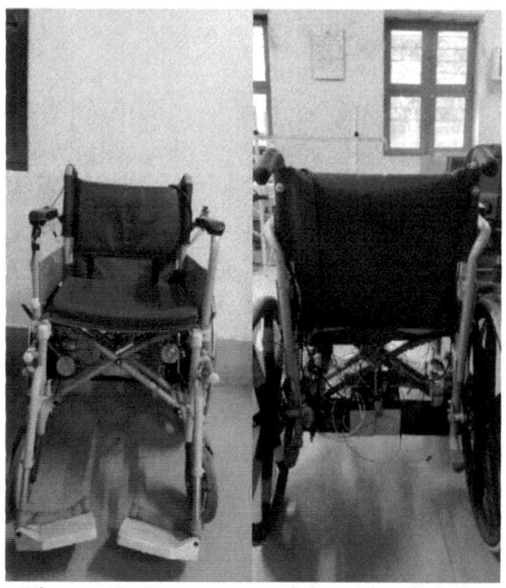

Figure 5. Prototype of the wheelchair.

4. Results

The aim of this project was to develop a wheelchair which would make the effort from the user for the movement of the wheel chair minimal. The wheelchair was tested in an indoor environment with 3 rooms and multiple obstacles placed in between. The movement of the wheelchair from Room 1 to Room 2 was executed successfully by the wheelchair avoiding all the obstacles in between. The time taken as compared to the manual control of the wheel chair was longer in automatic control as the motors are of lower power and the obstacle avoidance using ultrasonic is slightly slower to react. The results are tabulated in the table 1.

Table 1. Scenario 1 Results.

Source	Room1
Destination	Room 2
Obstacles	All obstacles avoided successfully
Time taken	3 Minutes
Time taken with manual control	2.5 Minutes

The movement from Room 2 to Room 3 was also achieved autonomously within 2 minutes. The wheel chair was not able to avoid one of the obstacles placed at a very low height, which required the user to adjust the position of the wheel chair using the joysticks. The time taken was 0.6 minutes more then what the user took while controlling the wheel chair manually
The results are tabulated in the table 2.

Table 2. Scenario 2 Results.

Source	Room2
Destination	Room 3
Obstacles	One obstacle not avoided
Time taken	2 Minutes
Time taken with manual control	1.4 Minutes

The movement from Room 3 to Room 1 was also achieved autonomously. But the time taken by the autonomous movement was almost a minute more than the manual control. This was attributed to the presence of more obstacles in the path making the autonomous movement slower. The results are tabulated in the table 3.

Table 3. Scenario 3 Results.

Source	Room3
Destination	Room 1
Obstacles	All obstacles avoided
Time taken	2 Minutes
Time taken with manual control	1.2 Minutes

5. Conclusion

The autonomous wheelchair reduced the human intervention to minimal extent and made the user independent of any caregivers for movement inside the home. The speed of movement though slower in the prototype presented in the paper can be improved with higher power motors. The obstacle detection and avoidance were also successful in avoiding and navigating the wheelchair but it can be improved with the aid of image-based systems but at the cost of higher system price.

6. References

[1] Khande R and Rajapurkar S 2022 Smart Voice and Gesture Controlled Wheel Chair *2022 6th International Conference on Trends in Electronics and Informatics, ICOEI 2022 – Proceedings* p 21776663

[2] Priyatharshini R, Muthu Selvam S, Kausic Narayanan M N and Kumar M A 2021 A Voice Controlled and Vision based Smart Wheel Chair for Paralyzed People *2021 2nd Global Conference for Advancement in Technology, GCAT* p 21417873

[3] Rahman Apu M A, Fahad I, Fattah S A and Shahnaz C 2019 Eye blink controlled low cost smart wheel chair aiding disabled people *IEEE Region 10 Humanitarian Technology Conference, R10-HTC* p 19472910

[4] Wanluk N, Visitsattapongse S, Juhong A and Pintavirooj C 2017 Smart wheelchair based on eye tracking *BMEiCON 2016 - 9th Biomedical Engineering International Conference* p 16692539

[5] HemaMalini B H, Supritha R C, Venkatesh Prasad N K, Vandana R and Yadav R 2020 Eye and Voice Controlled Wheel Chair *Proceedings of B-HTC 2020 - 1st IEEE Bangalore Humanitarian Technology Conference* p 20401894

[6] Khadilkar S U and Wagdarikar N 2015 Android phone controlled voice, gesture and touch screen operated smart wheelchair *2015 International Conference on Pervasive Computing: Advance Communication Technology and Application for Society, ICPC* p 15058392

[7] Park S, Ha T T T, Shivajirao J Y, Park J, Kim J and Hahn M 2013 Smart wheelchair control system using cloud-based mobile device *2013 International Conference on IT Convergence and Security, ICITCS* p 14047639

[8] Naresh S, Arunkumar R, Suriya I, Vinodh T and Radjaram B 2019 Design of powered wheelchair for a differently abled person *2019 IEEE International Conference on System, Computation, Automation and Networking, ICSCAN* p 19082837

[9] Jape P R and Jape S R 2018 Virtual GPS Guided Autonomous Wheel Chair or Vehicle *2018 3rd International Conference for Convergence in Technology, I2CT* p 18233190

[10] Farheen N, Jaman G G and Schoen M P 2022 Object Detection and Navigation Strategy for Obstacle Avoidance Applied to Autonomous Wheel Chair Driving *2022 Intermountain Engineering, Technology and Computing, IETC* p 21799423

[11] Ouiguini R and Belloulou B 1996 Navigation of an autonomous wheelchair in a structured environment *IECON Proceedings (Industrial Electronics Conference)* p 5533778

[12] Lee S, Lee N, Ahn J, Kim J, Moon B, Jung S H and Han D 2017 Construction of an indoor positioning system for home IoT applications *IEEE International Conference on Communications* p 17065665

[13] Salter J, Li B, Woo D, Dempster A G and Rizos C 2008 802.11 positioning in the home *2008 5th IEEE Consumer Communications and Networking Conference, CCNC* p 9811177

[14] Cooper A and Hegde P 2018 An indoor positioning system facilitated by computer vision *2016 IEEE MIT Undergraduate Research Technology Conference, URTC 2016* p 17595183

[15] Chen L and Cao Z 2018 Concatenating road take me home: Indoor navigation without infrastructure support *Proceedings of the International Conference on Parallel and Distributed Systems - ICPADS* p 17805370

[16] Miyamoto S, Hirayama Y and Morinaga N 1998 Indoor wireless local area network system using infrared and radio communications *5th Asia-Pacific Conference Communications and 4th Optoloelectronics Communications Conference: The Conference-Vitality to New Century, APCC/OECC 1999* p 6596272

IOP Conf. Series: Materials Science and Engineering 1295 (2023) 012004 doi:10.1088/1757-899X/1295/1/012004

Predictive analysis of Low power DC loads in Residential Buildings

Satish BA[1*], and P Usha[1]

[1] Department of Electrical and Electronics Engineering, Dayananda Sagar College of Engineering, Bengaluru, Karnataka, India

satish.eee@dayanandasagar.edu

Abstract. There has been a constant rise in use of loads powered from direct current (DC) among the residential loads. In the traditional Alternating Current (AC) grid voltage undergoes conversion multiple times before it reaches the loads requiring DC, which leads to substantial power loss in the converters. Implementation of a DC grid in the residential buildings can reduce the losses caused due to the conversions. The paper presents a prediction model for domestic low power DC loads which can be powered from a battery. The data of usage of the load as well as the data of the parameters on which the load usage is dependent on was collected over a period of a year. This data was fed to a Random forest-based machine learning algorithm to predict the usage of the loads in future. The state of charge (SOC) of the battery used to power the loads was compared with the load prediction to decide whether the load can be powered from the available power in the battery or not. This allowed the system to decide in real time on the source to be used to power the DC loads and providing the users a choice to change the load usage time in order to prevent power from being drawn from the AC grid.

1. Introduction

The rise in use of electronics as well as battery powered loads have led to an increase in the use of loads which run on Direct Current (DC). According to the report [1] more than 70% of the current residential loads can be powered using DC. The residential buildings receive power in the form of AC from the grid and this is rectified and converted to the DC voltages required for the respective loads. Figure 1 depicts the various conversions that takes place in an AC grid when powering a DC load. The PV panel on the rooftop produces DC voltage which is inverted to AC and then again converted back to DC to be stored in the battery. The incoming AC from the grid is rectified and fed to the battery. The DC voltage of the battery is inverted to AC which is supplied to all the loads. In case of DC loads the AC received is rectified and converted to the required DC voltage and then fed to the loads.

The losses due to conversion by using an AC grid to power DC loads has been discussed and compared in [2] with the conclusion that the use of DC grids leads to much lower losses as compared to AC grids. To reduce the number of conversions required, adoption of DC grid to power the DC loads is being researched on. The benefits of using DC appliances and the DC distribution have been discussed in [3], bringing out the need for changes in the market for effective adoption of DC loads and DC grids The use of DC loads with a DC grid also reduces the overall power consumption by the loads as studied in [4]. [5] presents the various components of a circuit that are redundant when DC power is used to power the DC loads, concluding that DC grids for DC loads will be economical as well as efficient as compared to the use of AC grids.

The adoption of DC grid is also beneficial for integration of solar photovoltaic panels into the domestic grid. [6] studies the various losses that happen in converters when PV panels are connected to the AC grid. All these losses can be avoided if the DC power produced by the PV panels is used to power the DC loads. Implementing a DC grid for efficient use of PV panels has been discussed in [7]. The benefits of a solar powered DC microgrid over an AC grid has been presented in [8]. DC voltages of different levels 380V,220V and 48V have been proposed as the DC grid voltages for residential applications. These various levels have been compared with respect to their efficiencies in [9] and it has been concluded by the author that a 48V DC is the most efficient option. A similar study on choosing the DC bus voltage was carried out in [10], and the authors here too conclude that a 48V DC bus is the most suitable option for a residential DC grid.

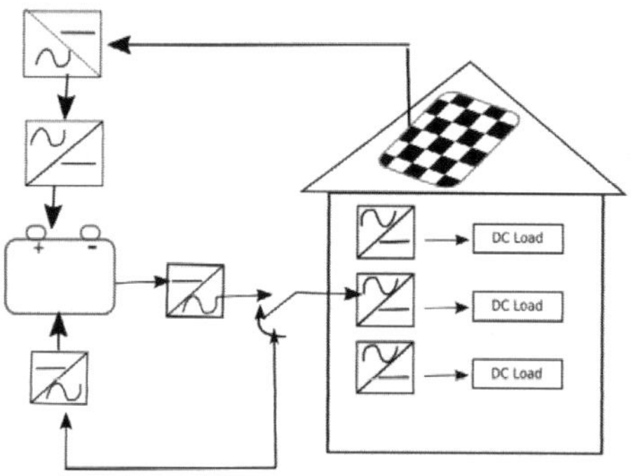

Figure 1. DC loads in traditional ac grid.

Considering the various advantages, a DC grid provides for powering the DC loads. Figure 2 depicts a model for connecting the DC loads directly to DC source with the aim to reduce the number of voltage conversions. The power from PV panel can be fed directly to the batteries with a DC converter. The power from the battery is fed to a DC bus which in turn is connected to the DC loads, which convert the voltage, if required, to their operating levels. Additional connection option to AC grid is provided in case the DC grid fails to provide the required amount of power to the DC loads.

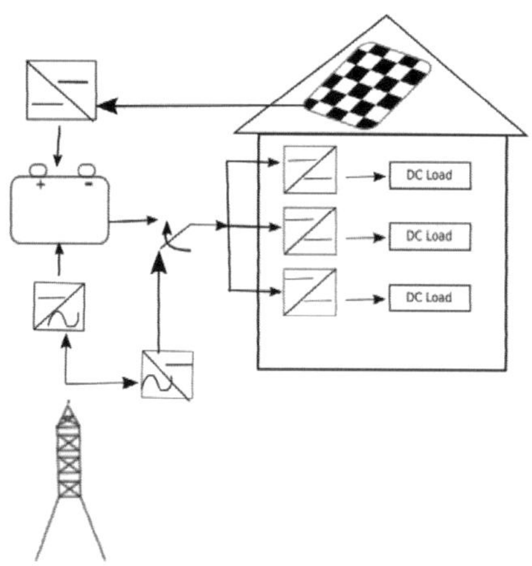

Figure 2. DC loads in DC grid.

To further reduce the number of voltage conversions required, this paper proposes a predictive power management system for a domestic DC grid. The aim of the proposed method is to ensure DC loads are powered directly from DC supply as much as possible, and minimal power is drawn from the AC grid for powering the DC loads. Figure 3 depicts the proposed model. The PV panel output is directly used to charge a battery without the intermediate conversion to AC. The loads are also connected directly to a DC bus being powered from the battery, eliminating another conversion used in an AC grid. Switches are provided between the AC grid and the Loads, as well as between AC grid and the battery. These switches are closed to connect the AC grid, to battery or to the loads only when, unlike the previous two configurations, the battery alone fails to provide the require power to the loads. Thus, this configuration uses 2 converters for a load, putting the other converters to use only when needed.

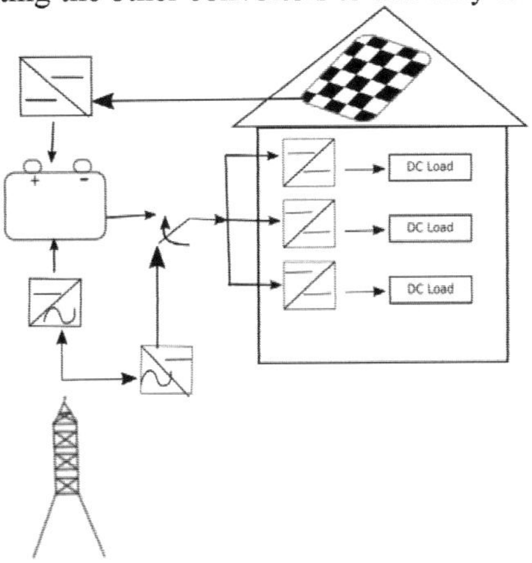

Figure 3. Proposed DC grid for DC loads.

Real time data of load usage was collected using an emulated setup consisting DC loads of various power levels. The loads were categorized based on their usage patterns. The data of the load usage was fed to a random tree generation algorithm to predict the usage of individual loads. Based on the prediction and the state of charge of the battery, the ability of the battery to support the load on a given day and at a given time was predicted. It was also demonstrated that in case the battery alone is unable to supply the required power, the battery can be charged from the AC grid or, the user can be given a choice shift the usage of the load to a different time when it can be powered from the battery. The availability of the granular prediction allows users to schedule and use loads efficiently, reducing the dependence on AC grid to power DC loads.

2. DC loads

Among the commonly used loads in a residential building, most of the loads readily have a DC alternative available and some are already in use extensively. The table 1 lists, along with their ratings, the most common, low power, DC loads that are used in a residential building.

Table 1. DC loads.

S. No.	Device	Voltage (V)	Current (max) (A)	Power (max) (W)
1	Mobile charger	5	0.5	2.5
2	Laptops	19	2-3	38-57
3	LED Bulb	9	0.5	4.5
4	LED Bulb	12	1.5	18
5	LED TV	24	3	72

Loads in a domestic setup generally follow a specific usage pattern and based on this, the loads can be categorized as shown below.

- Continuous use: Loads that are used very often for long duration of time, like the lighting loads of living rooms and porches. The time and duration of usage of these loads generally follow a specific pattern.

- Short term use: Loads which are used often but are for a short period of time like the lighting loads of rest rooms and closets. The short time of usage of these loads make them consume lesser energy compared to loads that are used for longer duration. The time and duration of usage of these loads are not as predictable as the loads that are used continuously.
- Rarely used: Loads that are rarely used like those of the attics. Analysing these loads for prediction will not be feasible as the loads do not add much to the daily energy requirement.

The duration and frequency of usage of loads depend on different factors the most important one being

- Time of the day: The usage of certain loads might be more during the day while others might show a higher usage during the night. Thus, considering the time of usage forms a important part in the analysis.
- Temperature: Loads used for cooling like fans and air conditioners are highly dependent on the temperature for their usage.
- Occupancy: Number of people in a room can affect amount of power drawn by the loads, for example, in case of the cooling loads the power required to cool a room would depend on how many people is it being occupied by.

The table 2 maps the loads in a residential building to the parameters on which they might depend on. The dependence of the usage on the predefined parameters makes the load usage predictable to a considerable extent.

Table 2. Load usage dependency.

Load	Time and Day	Temperature	Occupancy
Light	Yes	No	Yes
Fan	Yes	Yes	Yes
Mobiles	Yes	No	No
TV	Yes	No	No
Laptops	Yes	No	No

Machine learning is being used widely today to enable prediction in many fields including power systems and smart grids. Machine learning broadly can be stated as application of algorithms on the available data to be able to design a prediction model using statistical analysis (figure 4). The model in turn should be able to predict the output when provided with an input of the same nature as it was trained on. Various applications and advantages of machine learning in power systems have been reviewed in [11]. The application of machine learning in smart dispatch systems has been examined in [12]. Deep learning model for estimating losses in low voltage smart grids was studied in [13]. The use of rule based and machine learning model for demand response management of a building was presented in [14]. Aggregation strategies based on machine learning for short term load forecasting was studied in [15]. Machine learning has also been applied for very short-term load forecasting for factories in [16]. In this paper we have developed a machine learning model based on the random tree forest algorithm to predict the load usage at different time intervals and the same was compared with capacity of the battery to ensure that the battery, at a given state of charge, can supply the predicted power (figure 5).

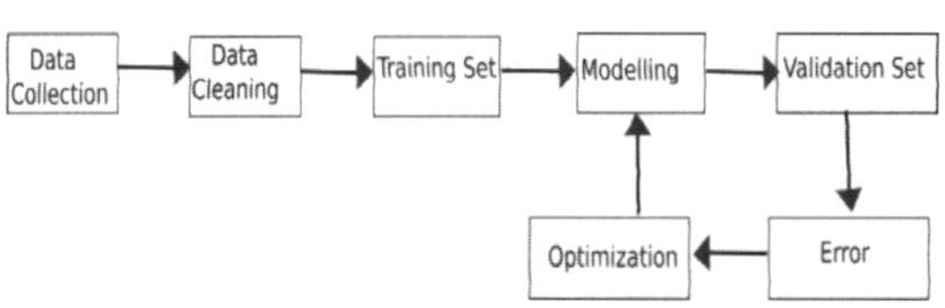

Figure 4. Model development.

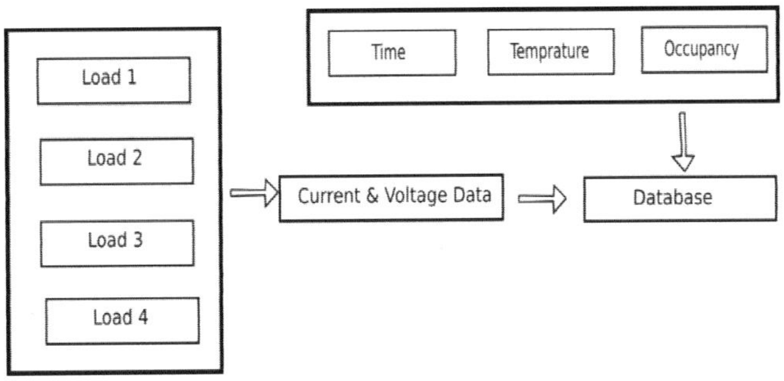

Figure 5. Data collection.

3. Methodology

3.1. Data Collection

Define abbreviations and acronyms the first time they are used in the text, even after they have been defined in the abstract. Abbreviations such as IEEE, SI, MKS, CGS, sc, dc, and rms do not have to be defined. Do not use abbreviations in the title or heads unless they are unavoidable. Having the right data and enough amount of it, is of primary importance for any prediction model to be able to generate accurate predictions. A setup comprising of four DC loads was created to enable real time collection of all the required data. The figure 5 depicts the block diagram of the system used for data collection. The system consisted of 4 loads connected to respective sensors, which in turn were connected to a local database for storage of the readings.

Out of the four loads used in the setup, three were made to be of continuous use type as these types of loads consumes bulk of the power in a residential building, and one was of short-term use type. Out of three continuous use type loads, usage of one load was made to depend only on the time of the day. The second one was made to depend on the time of the day as well as temperature where the power consumed by the load would increase with increase in temperature.

This was achieved by controlling the current delivered to the load using pulse width modulation based on the temperature. The third load was made to depend on the time of the day, temperature as well as occupancy. The power consumed would increase as occupancy increases or when the temperature increases. The variation of the current drawn by the loads with the change of temperature, as considered for the sample loads, is depicted in the table 3, and the variation with respect to the occupancy is depicted the table 4. These values are taken just to show the variation in the power consumed and do not depict the pattern followed by any practical load in real life.

The dependency of the loads on the parameters, as considered in the paper, is depicted in the table 5. The switching ON and switching OFF the load was controlled using relays which in turn were controlled using a micro controller. The randomness introduced by the controller while operating the loads ensured that the setup acted like a practical scenario.

Table 3. Current variation with temperature.

Temperature	Load Current
Less than 26 °C	0.5Imax
Between 26-30 °C	0.75Imax
Greater than °	Imax

Table 4. Current variation with occupancy.

Occupancy	Load Current
Greater than 4 People	Imax

Greater than 2 People	0.75Imax
Only one Person	0.5Imax

Table 5. Load categories.

Load No.	Category	Dependency
Load 1	Continuous Use	Time of Day
Load 2	Short Term use	Random
Load 3	Continuous use	Time of Day and Temperature
Load 4	Continuous use	Time, Occupancy and Temperature

Table 6. Parameters and sensors.

Parameter No.	Sensor	Interval
Date and Time	Real Time Clock	2 minutes
Current	ACS 712	2 minutes
Voltage	Voltage Divider	2 minutes
Temperature	LM 35	2 minutes

To collect the data on, time of usage of the loads, the current consumed by a load, input voltage, and the ambient temperature, at regular intervals of 2 minutes, the sensors listed in the table 6 were used. Readings from the sensors was collected by the controller and stored in a local database. The data collection from the loads was stored in the database only if the load was in use for more than four minutes. This ensured that the transients did not affect the predictions as they would not be added to the training data. Also, very short-term load usage is not suitable for predictions. The data collected from the sensors over a period was used as the input data for the prediction model. The figure 6 depicts setup used to emulate the DC load operation. As discussed before, it consists of four DC loads along with the sensors, the control, and power circuit. The setup has been operational for a year, providing real time data of the load usage and the dependent parameters which can be fed as input to train the predictive model. The data of the load usage was emulated locally instead of making use of the readily available data sets because an emulated setup allowed us to track the performance of the algorithm in real time by comparing the predicted values and the actual output obtained. The data collected over a year, at the intervals mentioned, was used as the data set to train the machine learning model. But before the data could be used by the model, any noise present in the data had to be removed by the data cleaning process.

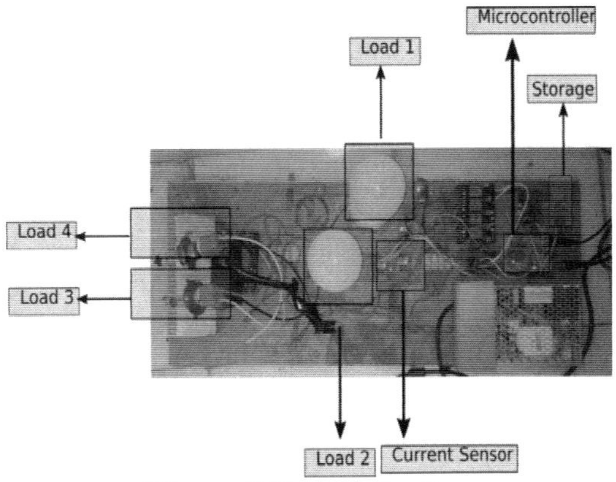

Figure 6. Data collection.

3.2. Model fitting and hyper parameter tuning

The cleaned data was fed as input to train the predictive model based on the random forest algorithm. The model was initialized with parameters based on general understanding of the system. The accuracy obtained from the training set was used to fine tune the hyper parameters, to obtain the best possible accuracy in the training set followed by best possible accuracy in the test set. The output of the model was a prediction value providing the likelihood of the load usage at a given time.

Random forest is a tree-based ensemble algorithm as proposed in [17]. The algorithm starts with a root node of the tree and creates several trees, each one independent of the other. The final prediction is arrived at by taking the mean of all predictions, of all the trees, as depicted in figure 7.

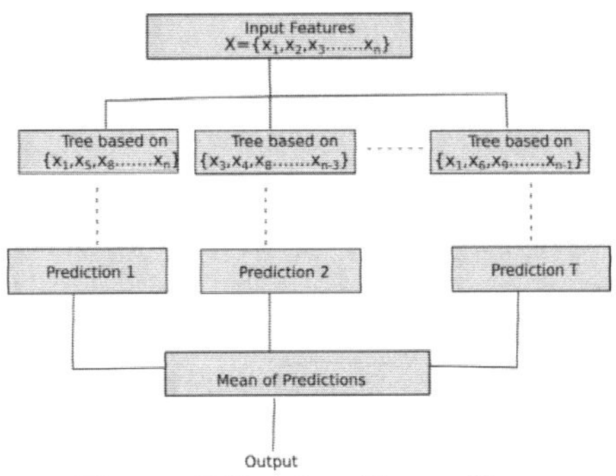

Figure 7. Random Forest Tree.

The creation of many trees gives the algorithm the name of forest, and the creation of trees is done in a random fashion, each tree taking a random set of features to create the tree from the given input features. The randomness of the algorithm has proven to be effective in negating the effects of noise and outliers in the data [18]. Regression trees have been found to be a lot more efficient than decision trees. In decision trees the splitting of the tree into different branches happens based on specific criterion and the final output is based on only one tree, which is not the case with random forest algorithms. The effectiveness of Random forest algorithms can be understood by looking at the various other scenarios where it has been applied successfully. In [19] random was applied to achieve better mapping of solar irradiance at higher altitudes. The possibility of 100% usage of renewable energy in a microgrid was explored using random forest algorithm in [20]. The features which will be treated as input to the nodes are the parameters on which the load usage is dependent. The output obtained from the prediction model, once implemented, will be the probability of a load being used at a given time. The values of probability vary between 0 and 1. The prediction value 1 indicating that the load will be surely in use for the given parameters, at the given time. Whereas a 0 indicating that the load will surely not be used for the given parameters, at the given time. The actual output values lie as floating-point number between 0 and 1. The paper treats the probability values of greater than 0.7 as the load being in ON state and values less than 0.7 as indication of load not being used at a given time. The value of 0.7 was arrived at with trial and error.

3.3. Prediction model

The random forest model was implemented using the Keras library in python. The data collected from the setup was split into 80% data for training and 20% data for testing of the model. The model was implemented to provide predictions of load usage for three different time intervals.

- Day prediction: After every time unit, the number of minutes the load might be used during the rest of the day was predicted.
- Hour Prediction: After every time unit, the number of minutes' load might be used in the remaining part of the hour was predicted.
- Short term prediction: The possibility of load being used within the next few minutes of the current time.

The accuracy of each of these three-time interval predictions, for all the 4 loads, were calculated by comparing the predictions with the actual load behaviour. The actual load usage and the predicted load usage, for each load, was saved in a database. The state of the load at given time was compared in both the scenarios to arrive at the conclusion whether the prediction was correct or not. Taking the average of all the comparisons provided the accuracy of the predictions. The steps used for finding the accuracy is depicted in the figure 8.

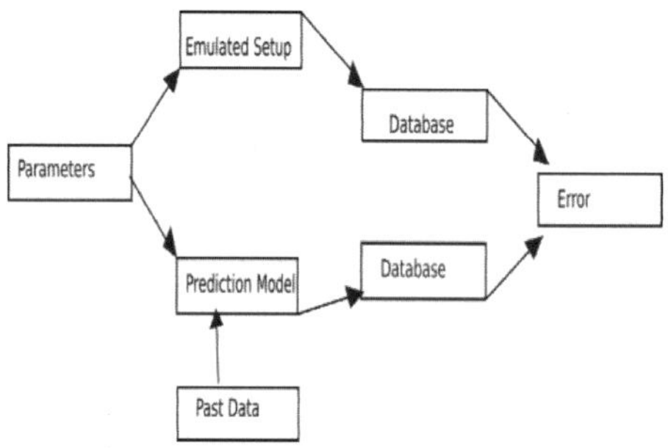

Figure 8. Steps for finding error.

3.4. Complete system

The figure 9 depicts the block diagram of the complete system. The database holds the past data of load usage along with the sensor outputs collected. This data is fed as input to the machine learning model. The output predictions of the machine learning model are sent as input to the control unit. The control unit combines the prediction value with the data of the State of Charge of the battery (SoC), and decides on sending the appropriate signals to the relays for powering the loads.

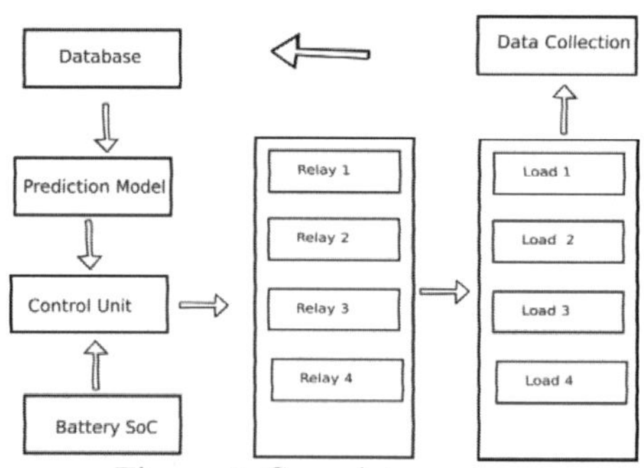

Figure 9. Complete system.

The algorithm in the control unit will aim to ensure that the loads be powered by battery for the maximum duration of time, minimizing the conversions required. The load usage data continuously gets updated in the database allowing the parameters of the prediction model to be modified according to the change in the usage patterns of the user.

4. Results and discussion

The prediction algorithm was run on a raspberry pi 4 board. The predictions of the algorithm were stored in a database at the time interval of every 10 minutes. The actual load usage
was also stored in a database at the same time interval. The above setup was allowed to run independently for a period of one week and at the end of the week the output obtained from the predictions and the actual behaviour was compared.

4.1. Accuracy of predictions

After the text edit has been completed, the paper is ready for the template. Duplicate the template file by using the Save As command, and use the naming convention prescribed by your conference for the name of your paper. In this newly created file, highlight all the contents and import your prepared text file. You are now ready to style your paper; use the scroll down window on the left of the MS Word Formatting toolbar.
The prediction algorithm was run on a raspberry pi 4 board. The predictions of the algorithm were stored in a database at the time interval of every 10 minutes. The actual load usage was also stored in a database at the same time interval. The above setup was allowed to run independently for a period of one week and at the end of the week the output obtained from the predictions and the actual behaviour was compared. The accuracy obtained during the training phase and testing is shown in the table 7, which was obtained by using the hyper parameters listed in the table 8 The mean error in predictions in number of minutes for predictions of the load usage for the rest of the day is given in table 9.

Table 7. Mean square error.

Load No.	Training set error	Test set error
Load 1	0.0008	0.006
Load 2	0.002	0.015
Load 3	0.0015	0.011
Load 4	0.003	0.023

Table 8. Hyper parameters.

Parameter	Value
Features	All Features
Trees	100
Elements	2
Elements for node	1
Depth	Until all leaf's are pure

Table 9. Mean error in number of minutes in an hour.

Parameter	Value
Load 1	3.06
Load 2	6.04

Load 3	5.36
Load 4	6.48

The table 10 gives the mean error in predictions in the number of minutes for the usage of load in an hour and, the table 11 gives in percentage, the error in the predictions of load usage. The comparison, for a complete day, between predictions and the actual usage of Load 1, which emulates the loads used continuously for long duration is shown in figure 10. The graphs indicate that the prediction follows the actual usage very closely. The mean error for one day was found to be 14 minutes for average full day use of 332 minutes.

Table 10. Mean error in number of minutes in an hour.

Load No.	Percentage
Load 1	14.58
Load 2	24.14
Load 3	36.38
Load 4	38.97

Table 11. Percentage of mismatch.

Load No.	Percentage
Load 1	6
Load 2	20
Load 3	10
Load 4	13

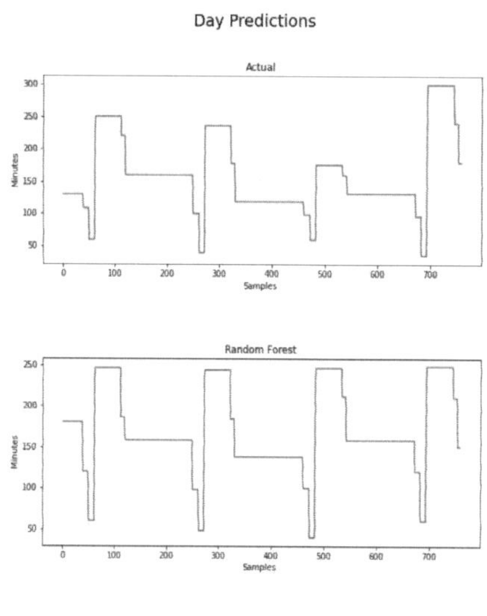

Figure 10. Day prediction.

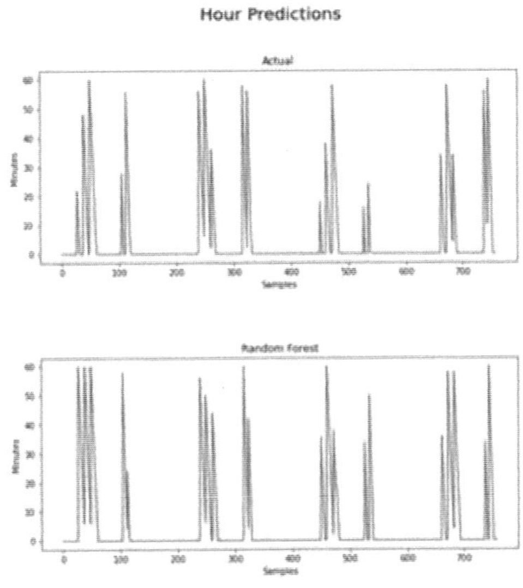

Figure 11. Hour prediction.

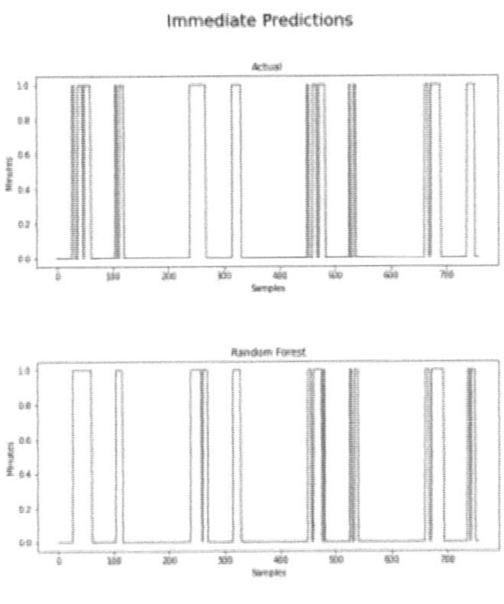

Figure 12. Load 1-minute predictions.

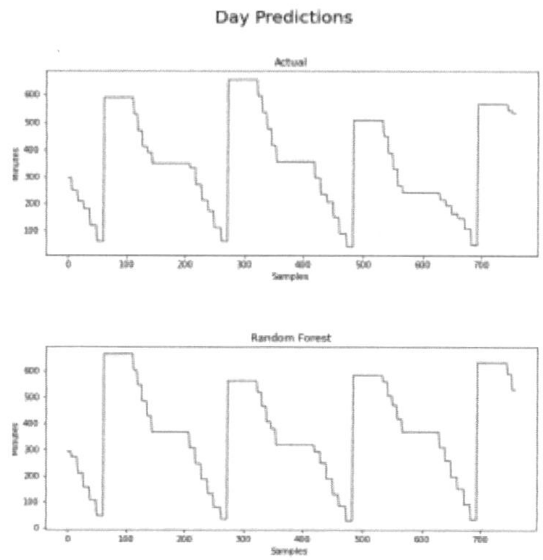

Figure 13. Load 2-day predictions.

In the case of the predictions within an hour, the mean error was found to be 6.04 minutes, the graphs for the same is shown in figure 11. The immediate prediction of the usage of the load was also pretty accurate with the error in predictions being only 6% as depicted in the figure 12. Load 2 emulates the loads which are used for short duration with no specific defined usage patterns. The random nature of these loads makes them difficult to predict accurately. Though cumulatively over a full day difference between the prediction and actual number of minutes the load gets used was found to be only 24.14 minutes, for full day usage of 250 minutes. The graph for which is plotted in figure 13. The error in predictions of usage in number of minutes with in an hour was found to be 6 minutes, which is relatively high as these loads get used for short duration of time in the order of 15 to 20 minutes. The comparison is plotted in the figure 14. In the case of immediate predictions, the randomness of the usage makes the prediction very poor with the system failing almost 20% of the time as shown in figure 15.

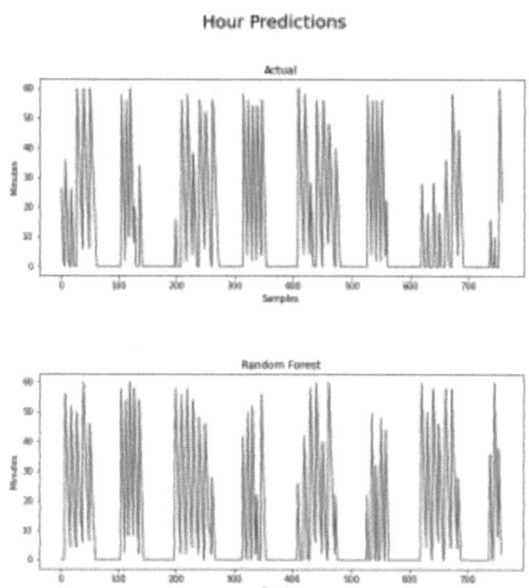

Figure 14. Load 2-hour predictions.

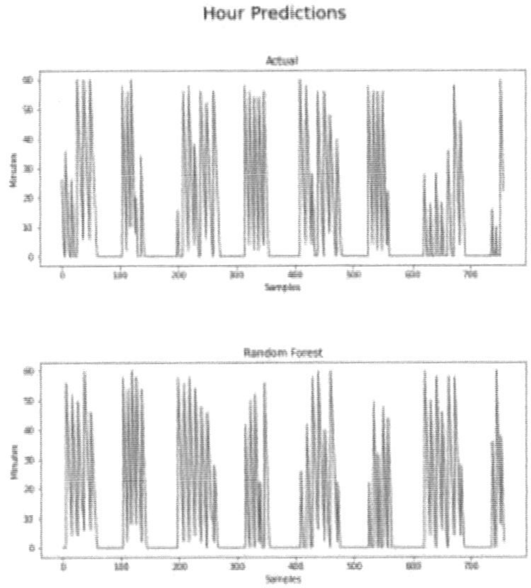

Figure 15 Load 2-minute predictions.

Load 3 emulates the loads whose usage depends on temperature, as well as on time. The usage is of continuous nature with the power drawn by the load varying with the variation of temperature. The error in prediction in number of minutes for a full day was found to be 34 minutes. The predictions for the usage within an hour had an error of 5 minutes on an average. The immediate predictions of load usage had an error of 10%. The increase in the error values can be attributed to the additional dependence on temperature.

Load 4 emulates the load whose usage depends on, temperature, occupancy, and the time. The power required to drive the load increases as the temperature or occupancy increases. The predictions for the full day usage had an error of 40 minutes with an average usage of the load being 600 minutes in a day. The mean difference in the predicted and the actual usage in an hour was found to be 10 minutes. Immediate usage predictions had an error of 15%, the lower accuracy can be attributed to the dependence on varying temperature as well as the randomness in the occupancy.

4.2. Control unit

The predictions of load usage were used to decide whether the loads can be powered from the charge available in the battery or not. The control unit accepted the output from the prediction algorithm, which was the number of minutes each load is expected to be used in the remaining day and compared this with the SoC of the battery to decide if the battery can supply the required power to the load or not. The ratings of the loads that were considered in the study are listed in the table 12. The table 13 depicts the decisions that the control unit makes based on a sample input received. The total Ampere-hour (Ah) required for a day according the prediction is given to be 29Ah. Because the requirements of some loads might be more during the transient periods, a tolerance of 10% was added to predictions. Thus, if the prediction was 29Ah the requirement would be 33AH. To satisfy this a 36AH battery was considered which can supply power to all the loads during the full day if it is charged to 100%. But if the charge at the beginning of the day is at 75%, only Load 1, Load 2, and Load 4 can be supplied for the full day. The choice of loads is based on which loads can be supplied for the longest duration of time with the given state of charge. Similarly with the State of Charge at 50%, only loads 2 and 4 will be powered from the battery. As of now the study did not consider the case of battery being charged while it is supplying the power. The table 14 describes the case in which the predicted user requirement for the next hour exceeds the power that the battery can supply, which can be avoided if the user can distribute the load usage and allowing the battery to be charged for the next hour. The predicted value of AH usage for the present hour was 4.55 Ah and the next hour was 1.2Ah.

Table 12. Current Ratings.

Load No.	Percentage
Load 1	1
Load 2	0.5
Load 3	3
Load 4	2

Table 13. Sample predictions for the day.

Load No.	No. of hours of use	Ah required
1	4	4
2	2	1
3	4	12
4	6	12
Total	29AH	
Battery Capacity	30AH	
SoC	Loads connected to battery	
100%	All loads	
75%	Loads 1,2, and 4	
50%	Loads 2 and 4	

Table 14. Load rescheduling suggestion.

Load No.	Use in Current hour (hours)	Use in Next hour (hours)
1	0.5	0.2
2	0.1	0
3	0.8	0.2
4	0.8	0.2
Total (AH)	4.55 Ah	1.2 Ah
Current Battery Capacity	30Ah	
Suggested Usage by Algorithm		
1	0.5	0.2
2	0.1	0
3	0	1.0
4	0.8	0.2
Total	2.95 Ah	2.8 Ah

With the current capacity of battery at 4Ah only, because of its state of charge, one of the loads had to be connected to the grid. On the other hand, if the user is suggested to shift the usage of Load 3 to the

next hour, not sacrificing the duration of use, then the Ah requirements change to 2.95 Ah and 2.8Ah, which the battery will be able to supply with minimal charging from the PV. Thus, we could avoid the usage of an extra converter which would have been used if the load was connected to the grid.
The control unit was implemented using Raspberry pi 4 and Arduino uno boards. The Arduino was used to read the state of charge of the battery and store the same in shared database. The state of charge measured and stored by Arduino, was accessed by the Raspberry pi board, which was also executing the prediction algorithm. The control algorithm executing in the Raspberry pi board combined the output from the prediction algorithm and the state of charge read from the database to make the decision on powering the loads from the battery based on the which, signals were sent to the relays to connect the required power source of the loads. Figure 16 shows the hardware implemented to realize the control unit with 4 loads. This granular prediction for every load allows the optimal use of batteries, to ensure that maximum number of DC loads are powered using DC supply avoiding the various conversions required in an AC grid [21].

5. Comparison

The advantage of using a DC converter over an AC converter has been studied in detail in [21] with the conclusion that low power DC boost converters have almost 3 times lesser loss compared to the AC converters. Thus, clearly decreasing the use number of AC converters used for a load to be powered will increase the overall efficiency. To show how the number of converters required in the proposed model will be lesser than a traditional AC grid, let us look at a comparison between the three types of grids.

- Type 1: DC Loads in traditional AC grid.
- Type 2: DC loads in a DC grid.
- Type 3: DC loads in a DC grid with load-based predictions as proposed in the paper.

5.1. Scenario 1: Battery has charge to supply to all the loads

Table 15 compares the number of converters use in case the battery has enough charge in it to supply power to all the loads. In the traditional AC grid 4 converters will be in use always. In the case of a DC grid with no prediction algorithm, three converters will have to be used. In the case of a DC grid with prediction algorithm, if the algorithm predicts that battery can power the loads depending on the load predictions only 2 converters will be needed one for charging the battery from the solar panels and the second used to adapt the voltage as needed for the loads. Clearly reducing the number of voltage conversions needed per load as compared to the traditional AC grid.

Table 15. Comparison scenario 1.

Grid type	Converter type	Number of converters
Type 1	AC-DC	2
	DC-AC	2
Type 2	DC-DC	3
Type 3	DC-DC	2

5.2. Scenario 2: Battery can supply only partial loads

Table 16 shows the scenario in which the battery is not enough to power all the loads. In the case of a traditional AC grid the grid would be charging the battery for supplying the extra power needed. In a DC grid without prediction, Three converters will be needed, one from the panels, one from the grid, and one for the individual loads.

Table 16. Comparison scenario 2.

Grid type	Converter type	Number of converters
Type 1	AC-DC	2
	DC-AC	2
Type 2	DC-DC	3
Type 3	DC-DC	2

In the case of DC grid with prediction algorithm, we will have two options. One we can connect the battery to the grid to charge the battery based on the need of the load or the user can be prompted to shift the load usage to a time duration where the load requirements are such that battery might be able to support the loads. By choosing to shift the load usage time, we can still manage to operate with only two converters in use for a load.

Thus, in both the cases the proposed model requires lesser number of converters than the traditional AC grid or a DC grid with no prediction algorithm. The reduction in use of the number of converters will in turn reduce the losses per load in the overall grid, thus leading to a more efficient usage of the renewable energy resources.

6. Conclusion

The paper proposed a predictive model for the DC loads by categorizing the loads based on their usage. The model provided prediction of load usage, in number of minutes, during the rest of the day, rest of the hour, from a given time as well as the probability of the load being used at a given time. The predictions were accurate for loads which had more predictive patterns, as compared to loads which were used randomly. It was also demonstrated how predictions of the load usage can be used to decide on source used to power the loads, based on the state of charge of the battery. It was shown that the number of converters required to power the DC loads are minimum in the case of DC grids with predictive algorithm.

Such a predictive system implemented in a real time system can make DC grids more efficient and hence lead to a wider adoption of the same. A DC grid in turn will make use of renewable energy source like the solar panels more efficient, hence increase the adoption of renewable energy sources in the residential buildings.

7. References

[1] Starke M, Tolbert L M and Ozpineci B 2008 AC vs. DC distribution: A loss comparison *Transmission and Distribution Exposition Conference: 2008 IEEE PES Powering Toward the Future, PIMS* p 10024869

[2] Kumar S, Yashaswini H K, Sharma N K, Bajaj M, Naithani D and Maindola M 2022 Classical Secondary Control Techniques in Microgrid Systems - A Review *Proceedings - 2022 2nd International Conference on Innovative Sustainable Computational Technologies, CISCT* p 22682768

[3] Vossos V, Pantano S, Heard R and Brown R 2017 DC Appliances and DC Power Distribution A Bridge to the Future Net Zero Energy Homes *Eedal* **2017** 1

[4] Sanjeev P, Padhy N P and Agarwal P 2015 DC grid initiative in India *IFAC-PapersOnLine* **48** 114

[5] Wunder B, Ott L, Szpek M, Boeke U and Weiß R 2014 Energy efficient DC-grids for commercial buildings *INTELEC, International Telecommunications Energy Conference (Proceedings)* p 14790960

[6] Rachmildha T D, Fadel M and Haroen Y 2019 Study on Conversion Losses of Several Converter Topologies Used in Grid Connected Photovoltaic System *IFAC-PapersOnLine* **52** 461

[7] Rauf S, Wahab A, Rizwan M, Rasool S and Khan N 2016 Application of dc-grid for Efficient use of solar PV System in Smart Grid *Procedia Computer Science* **83** 902
[8] Gerber D L, Vossos V, Feng W, Marnay C, Nordman B and Brown R 2018 A simulation-based efficiency comparison of AC and DC power distribution networks in commercial buildings *Appl Energy* **210** 1167
[9] Siraj K and Khan H A 2020 DC distribution for residential power networks—A framework to analyze the impact of voltage levels on energy efficiency *Energy Reports* **6** 944
[10] Moussa S, Ghorbal M J Ben and Slama-Belkhodja I 2019 Bus voltage level choice for standalone residential DC nanogrid *Sustain Cities Soc* **46** 101431
[11] Ibrahim M S, Dong W and Yang Q 2020 Machine learning driven smart electric power systems: Current trends and new perspectives *Appl Energy* **272** 115237
[12] Yin L, Gao Q, Zhao L, Zhang B, Wang T, Li S and Liu H 2020 A review of machine learning for new generation smart dispatch in power systems *Eng Appl Artif Intell* **88** 103372
[13] Velasco J A, Amaris H and Alonso M 2020 Deep Learning loss model for large-scale low voltage smart grids *International Journal of Electrical Power and Energy Systems* **121** 106054
[14] Pallonetto F, De Rosa M, Milano F and Finn D P 2019 Demand response algorithms for smart-grid ready residential buildings using machine learning models *Appl Energy* **239** 1265
[15] Feng C and Zhang J 2020 Assessment of aggregation strategies for machine-learning based short-term load forecasting *Electric Power Systems Research* **184** 106304
[16] Walther J, Spanier D, Panten N and Abele E 2019 Very short-term load forecasting on factory level - A machine learning approach *Procedia CIRP* **80** 705
[17] Schonlau M and Zou R Y 2020 The random forest algorithm for statistical learning *Stata Journal* **20** 1
[18] Kumar S, Paliwal H, Vyas S, Sekhor S, Dave V and Rao S S 2021 Dynamic Wireless Power Transfer in Electric Vehicles *Journal of Physics: Conference Series* **1854** 012014
[19] Babar B, Luppino L T, Boström T and Anfinsen S N 2020 Random forest regression for improved mapping of solar irradiance at high latitudes *Solar Energy* **198** 81
[20] Kiptoo M K, Adewuyi O B, Lotfy M E, Ibrahimi A M and Senjyu T 2020 Harnessing demand-side management benefit towards achieving a 100% renewable energy microgrid *Energy Reports* **6** 680
[21] Gerber D L and Musavi F 2020 AC vs. DC Boost Converters: A Detailed Conduction Loss Comparison *2019 IEEE Third International Conference on DC Microgrids (ICDCM)* p 20197804

IOP Conf. Series: Materials Science and Engineering 1295 (2023) 012005 doi:10.1088/1757-899X/1295/1/012005

Development of a Surveillance Robot with a Rocker Bogie Design for Versatile Terrain Navigation

Sreevidya T R[1], Nishant Pathak[1], Sampurna Nand Shukla[1], Saurav Giri[1], and Yashovardhan[1*]

[1] Department of Electrical and Electronics Engineering, Dayananda Sagar College of Engineering, Bengaluru, Karnataka, India

1ds19ee104.yashovardhan@gmail.com

Abstract. This project focuses on the design and development of a multi-terrain robot based on the rocker bogie mechanism. The robot is equipped with a camera, GPS module, Wi-Fi module, ultrasonic sensor for obstacle detection, and gas sensors for detecting carbon monoxide, methane, smoke levels, and AQI. It utilizes internet connectivity to transmit data and presents the collected information through a website dashboard. The project explores the potential applications of the robot, including surveillance, environmental monitoring, search and rescue, industrial inspections, and agriculture. The integration of various sensors and connectivity capabilities allows for real-time monitoring and analysis of critical data in challenging terrains and environments.

1. Introduction

Multi-terrain robots are essential in diverse environments, known for their adaptability and mobility across various challenging terrains [1]. They excel in navigating rough surfaces, steep slopes, rocky landscapes, and obstacles, making them valuable in critical applications [1]. These robots require advanced mobility, adaptability, and durability, as they face various terrains and conditions. The rocker bogie suspension system, renowned for its success in NASA's Mars rovers, addresses these challenges effectively. This design combines wheels and articulated suspension components to offer stability, articulation, traction, manoeuvrability, and flexibility, enabling robots to navigate diverse terrains with ease [2]. By integrating this design into multi-terrain robots, such as surveillance robots, they become versatile assets across applications like planetary exploration, environmental monitoring, search and rescue missions, and industrial inspections.

In our project, we've developed a ground rover, referencing the proven rocker-bogie suspension system, renowned for its exceptional stability and obstacle navigation [2]. This rover is equipped with sensors for environmental monitoring, including temperature, humidity, and air quality, which enable real-time data collection and analysis. The analysis of various parameters, including temperature, pressure, humidity, air quality, carbon monoxide (CO), and methane (CH_4) levels, provides valuable insights across multiple domains [3]. These insights support environmental monitoring, health and safety assessments, weather forecasting, climate studies, agricultural management, energy efficiency optimization, urban planning, and infrastructure development.

Through careful analysis of these parameters, patterns, trends, and environmental variations can be identified, contributing to a better understanding of climate patterns, air pollution levels, and potential health risks. Furthermore, this analysis assists in evaluating the impact of human activities on the environment, optimizing resource management, enhancing agricultural practices, and making informed decisions related to energy consumption, urban development, and infrastructure design. In essence, the analysis of these parameters improves decision-making processes, encourages sustainable practices, and safeguards ecosystems and human well-being.

2. Conventional methodology

After conducting a thorough literature survey, we determined that hybrid robots, combining wheels and legs, are the most suitable for our project [4]. Therefore, our project will incorporate both a wheel-based locomotion system and a suspension system with legs (figure 1). This conclusion led us to adopt the Rocker Bogie design as the foundation for our robot's mobility capabilities as it contained both wheels and individual arms.

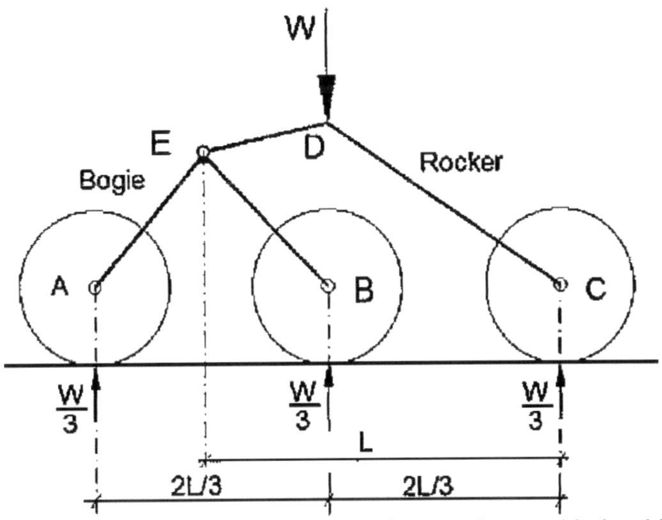

Figure 1. The block diagram of the proposed structure in accordance with the objectives defined for the robot.

The conventional Rocker Bogie system consists of a set of wheels and a series of articulated suspension components that work together to ensure all wheels maintain contact with the ground, even on uneven terrain. It is named after the two primary components: the rocker arm and the bogie. The rocker arm is a single beam that connects the robot's body or chassis to the bogie assembly. It acts as a pivot point for the suspension system, allowing the robot to maintain stability and keep all wheels in contact with the ground. The rocker arm adjusts its angle to accommodate changes in the terrain, ensuring the robot remains balanced. The bogie assembly comprises multiple wheels arranged in a parallel configuration. It is connected to the rocker arm through a set of axles, allowing independent movement of each wheel. The bogie assembly provides stability and distributes the weight evenly across the wheels, enhancing the robot's capability to traverse rough and uneven surfaces.

The Rocker Bogie design offers several key advantages:

- Stability: By distributing weight across multiple wheels and suspension arms, it ensures stability on uneven surfaces and reduces the risk of tipping.
- Articulation: Its independent wheel and suspension movement enables adaptation to varying terrain, efficiently overcoming obstacles.
- Enhanced Traction: Improved grip and weight distribution allow navigation on loose or slippery surfaces.
- Flexibility: Adapts to changes in terrain height or irregularities without compromising stability.

- Manoeuvrability: Enables tight turns and navigation through narrow spaces for efficient exploration and operations.

While widely used in space exploration, it has untapped potential in areas like search and rescue, industrial inspections, and environmental monitoring. We've adapted this design for terrestrial applications to harness its benefits (figure 2). The conventional rocker bogie structure comes in two forms:

- Linear-Type Bogie Assembly: Offers stability on rough surfaces with high friction, like cratered terrain, and provides high speed due to its aerodynamic structure.
- Trapezoidal-Type Bogie Assembly: Excels on hilly surfaces with high inclinations, such as those with protrusions, and delivers superior torque thanks to its stable structure.

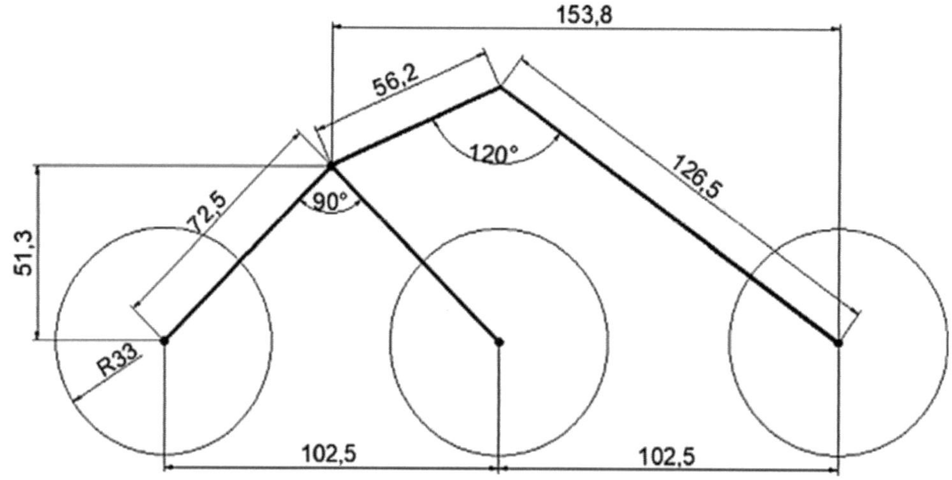

Figure 2. The block diagram of a trapezoidal type rocker bogie structure, studied to understand the rocker-bogie structure.

By seamlessly integrating both the linear-type and trapezoidal-type bogie assemblies, we have engineered a sophisticated rocker-bogie structure that leverages the unique advantages of each configuration. This hybrid design enhances the system's overall resilience and adaptability, making it exceptionally well-suited for navigating diverse terrains and challenging environments, thus offering a versatile and robust solution for a wide range of applications.

3. Design of proposed report

3.1. Design calculations (bogie link)

In this study, we base our calculations on the fundamental assumptions that underlie the Rocker Bogie mechanism. The mechanism itself is subject to specific construction and constraints, particularly the requirement that angles ABC and ADE must be equal to 90° [5]. To achieve a high degree of precision and accuracy in the design process, we opt for the utilization of AutoCAD software (figure 3).

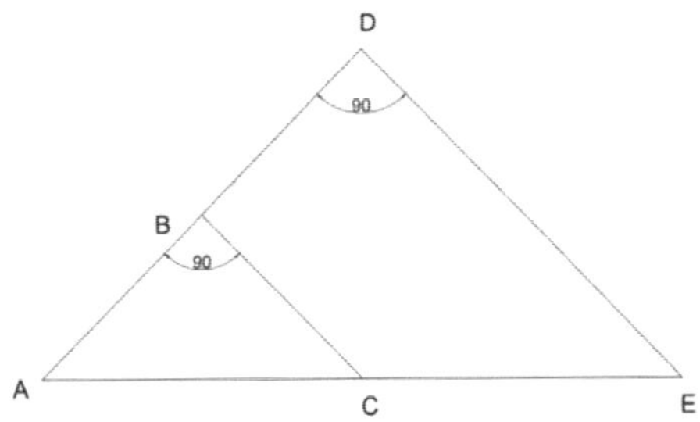

Figure 3. The general structure – the isosceles triangles in the suspension.

To ensure stability while navigating obstacles, it becomes necessary to raise one pair of wheels at a time. Therefore, our initial step involved determining the dimensions of the Bogie links.
We make the setup for calculation as first pair of wheels placed horizontally at the end of rising with 15mm extra space for safety purpose. on the same second pair is just before the start of raising with 15mm extra space for safety purpose (figure 4 and figure 5).

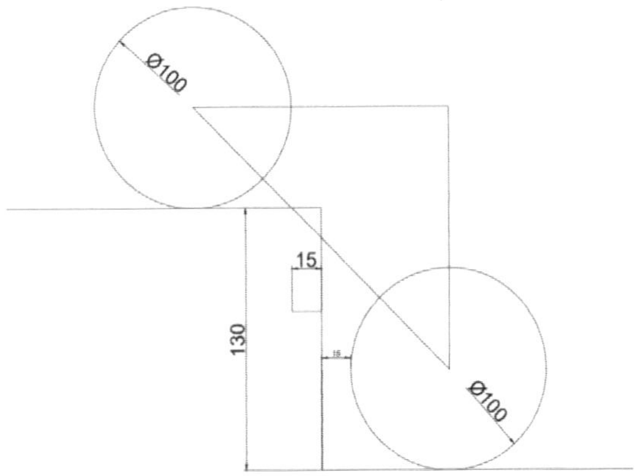

Figure 4. CAD diagram of the suspension in climbing a stair step of 130 mm (AB) size.

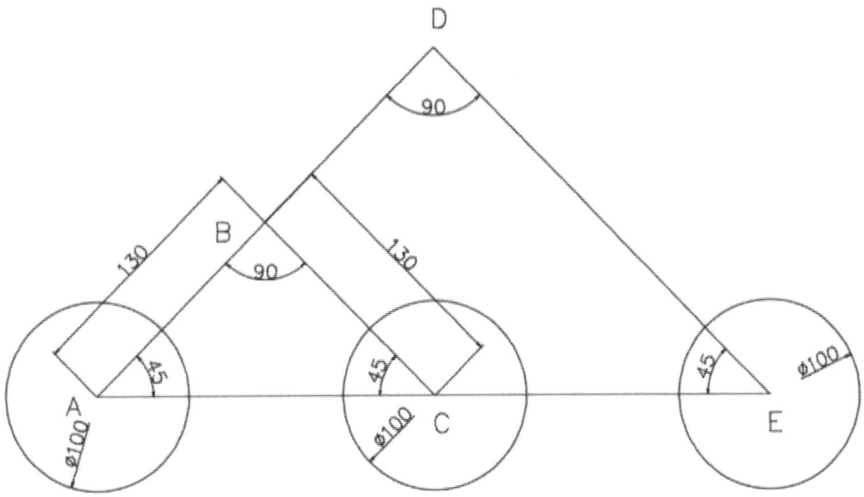

Figure 5. CAD diagram of the general structure with dimensions of bogie links specified.

3.2. Design calculations (rocker link)

To make space for the payload box, receiver, and master circuit, it is crucial to keep the top surface of the Rocker Bogie flat. To achieve this, we make a cut along the upper edge of Triangle Δ ABC, ensuring that Triangle Δ ABC's height surpasses the height of the cut edge from its base. To ascertain the length of the bisect line XY, we continue by forming Triangle Δ DXY, h = 99.49 mm, H = 182.847 mm, DE = 260 mm, AE = 367.69 mm.

3.3. Top-structure removal calculations

Based on the calculations and acquired data, we determine the dimensions of the Rocker Bogie links and the final structure obtained is in the figure 6 and figure 7.

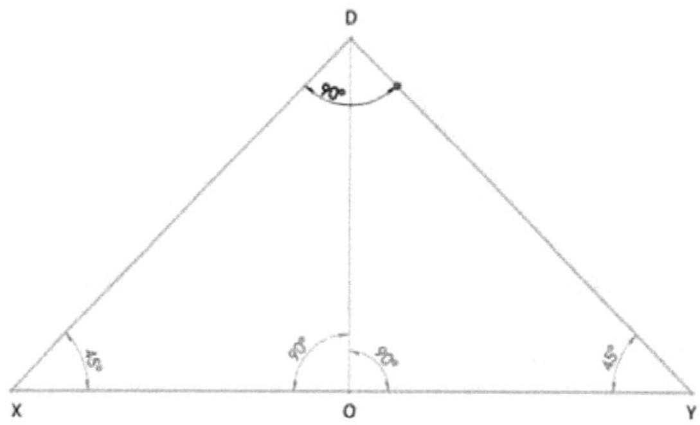

Figure 6. CAD Diagram of triangle to be removed from the original structure to compensate for accessories.

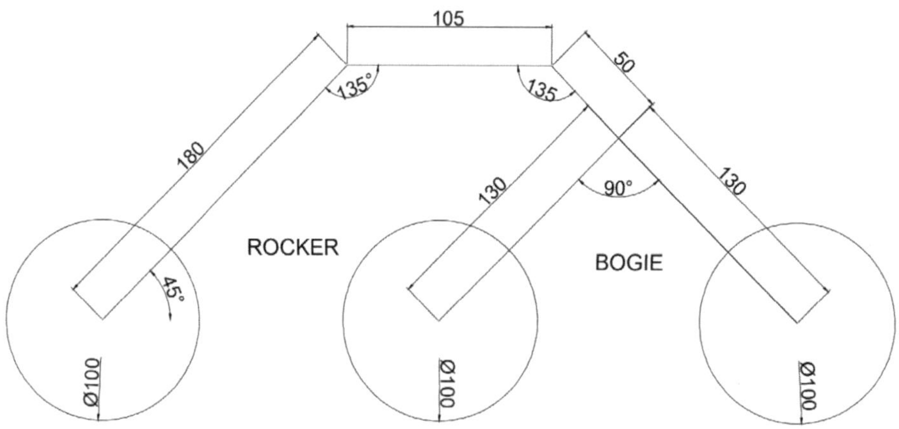

Figure 7. CAD Diagram of final suspension of the prototype with all dimensions specified.

These design calculations for the rocker bogie robot reveal its unique characteristics when compared to other designs. The primary dimensions, especially the minimal length of the bogie link (130mm) and the wheelbase (183.84mm), were derived to optimize stability and mobility. The calculations extend to the rocker link construction, where constraints dictate a length of 183.847mm.

By further trigonometric analysis, the critical height of both Triangle Δ ABC and Δ ADE was determined, ensuring the flat top surface necessary for accommodating additional components. Notably, the removal of the top structure was calculated precisely to achieve the desired outcome, with XY bisecting the system at approximately 105mm. This meticulous approach to design results in a rocker bogie robot uniquely tailored to its specific requirements, setting it apart from other designs in terms of its dimensions and adaptability.

4. Simulation

The designs obtained in Auto CAD are then transformed and simulated in SolidWorks software to create a line-diagram of a more stable structure (figure 8-10). The created design thus provides:

- high stability and aerodynamic properties for holes and craters, whereas also
- provides stability & torque while encountering obstacles and protrusions like stairs etc.

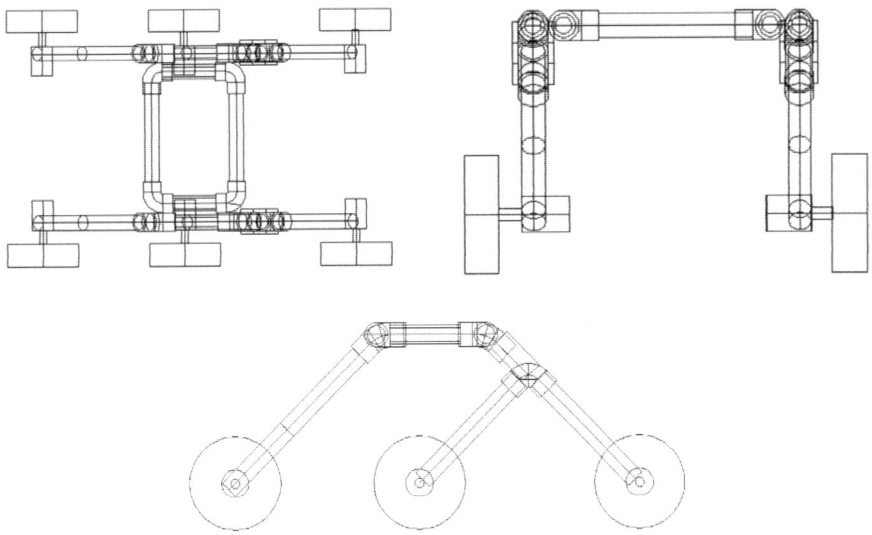

Figure 8. The top, front and side view line diagram of the designed structure of the Rocker -Bogie suspension.

These line-diagrams are then simulated into solid counterparts to study its integrity at the different joints of the robot.

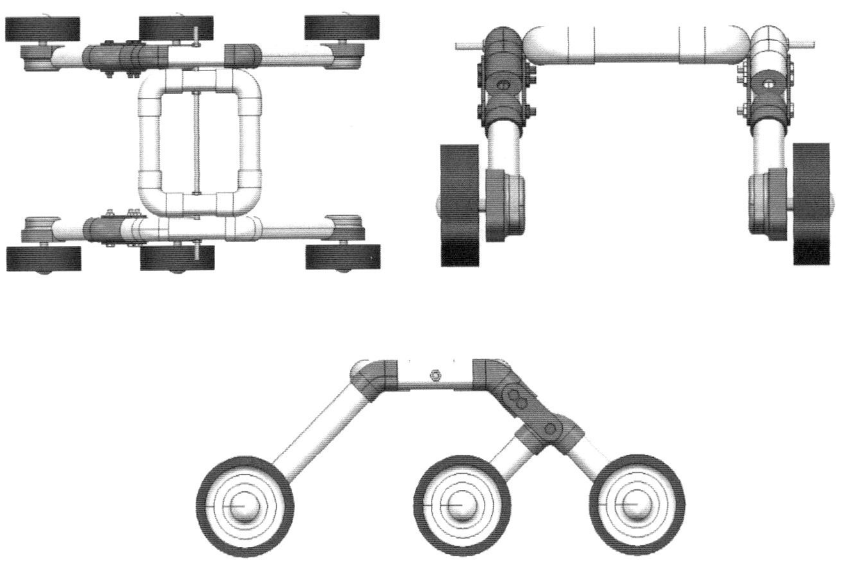

Figure 9. Side, top, and front view of the simulation of the specifically designed Rocker -Bogie Suspension in the SolidWorks software.

IOP Conf. Series: Materials Science and Engineering 1295 (2023) 012005 doi:10.1088/1757-899X/1295/1/012005

After multiple iterations of stability check and fine tuning, a final structure is created and which would stay upright in rough terrains and would be able to climb obstacles [6]. This structure is then modified into a more stable and robust structure which would also have the load carrying capacity.

Figure 10. Isometric view of the simulation of the more stable and robust Rocker -Bogie Suspension in the SolidWorks software.

5. Circuit connections

Robot connections are divided into three parts.

5.1. Transmitter circuit

The transmitter circuit consists of joysticks, a battery, an Arduino NANO, three toggle switches for light control and NRFL 2401 Radio Frequency RF Transceiver connected to the circuit using a regulator. The transceiver comes attached with an antenna [7]. The signal from the joystick is sent by the Arduino NANO to the receiver circuit through the transceiver for respective actions of control (figure 11).

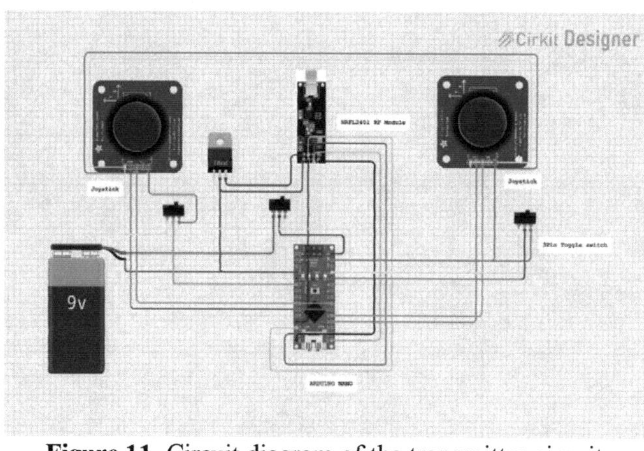

Figure 11. Circuit diagram of the transmitter circuit.

5.2. Receiver circuit

The receiver circuit is an integral part of the overall system and is primarily composed of an NRFL 2401 Radio Frequency (RF) Transceiver, which is connected to the master circuit. The RF transceiver is equipped with an antenna to facilitate wireless communication. The primary function of the receiver circuit is to capture the radio frequency signal transmitted by the remote controlling transmitter. This signal contains instructions and commands from the user, allowing them to remotely control the robot's movements and actions. The NRFL 2401 RF transceiver is responsible for receiving this signal and passing it along to the main micro-controller.

When the RF transceiver receives the radio frequency signal, it utilizes its antenna to capture the transmitted data. The transceiver then processes the received signal and extracts the encoded commands. These commands are subsequently forwarded to the main micro-controller of the system. The main micro-controller, such as the Arduino MEGA, is connected to the receiver circuit and serves as the central control unit [8]. It receives the commands from the RF transceiver and interprets them accordingly. Based on the received instructions, the micro-controller generates appropriate signals to control the robot's motors, sensors, and other components. By effectively relaying the radio frequency signal from the remote-controlling transmitter to the main micro-controller, the receiver circuit enables seamless communication and remote-control functionality. This ensures that the user's commands are accurately captured and executed by the robot, allowing for efficient operation and control over its movements and functionalities.

5.3. Master circuit

The master circuit consists of the main micro-controller, Arduino MEGA, which serves as the central control unit. It is responsible for coordinating and controlling the entire system. The micro-controller is connected to various components that enhance the functionality of the robot. Firstly, the Arduino board is connected to an ESP32 CAM module which enables the robot to transmit a live camera feed, allowing remote users to see the surroundings from the robot's perspective in real-time [9]. This feature enhances the robot's ability to navigate and interact with its environment effectively. A GY-NEO 6MV2 GPS Module is connected to the Arduino board. This module provides accurate positioning information by constantly tracking the robot's coordinates [10]. The GPS data is crucial for monitoring and controlling the robot's movements and can be utilized for mapping purposes or real-time tracking applications. In addition to the camera and GPS modules, the Arduino board is also connected to an ESP8266 Wi-Fi Module. This module facilitates the transmission of data gathered by various sensors and modules to a remote location or control station [11]. By leveraging Wi-Fi connectivity, the robot can send live data feeds, including camera footage, GPS coordinates, and other sensor readings, to be monitored and analyzed remotely.

The micro-controller is further connected to an ultrasonic sensor, which serves as part of an obstacle detection mechanism. This sensor emits sound waves and measures the time it takes for the waves to bounce back after hitting an object. By analysing the time taken for the sound waves to return, the robot can assess the presence of obstacles or hindrances in its path. If an obstacle is detected, the micro-controller can trigger appropriate actions or send warnings to prevent collisions or accidents.

To monitor the air quality in the surrounding environment, three sensors are employed: MQ-02, MQ-04, and MQ-07. These sensors are capable of detecting smoke levels, methane levels, and carbon monoxide levels respectively [11,12]. These gases can be harmful to living organisms, so monitoring their presence is essential for ensuring a safe environment. The sensor readings provide valuable data for assessing air quality and taking necessary measures to mitigate potential risks. Additionally, these gas sensors can be easily replaced with sensors designed to detect other gases if needed. The ranges mentioned in the table are typical data ranges, and the actual sensitivity and detection range of each sensor can be varied based on factors such as operating conditions and calibrated accordingly by changing the values of resistors used.

Table 1. Summary of the range of various gas sensors used.

Gas Sensor	Gas Detected	Typical Values	Typical Data Range
MQ – 02	Methane, Hydrogen, CO2, CO, Propane	0.1V to 4.0V	300 to 10000 ppm
MQ – 04	LPG (Propane +Butane)		300 to 10000 ppm

MQ – 07 Carbon Monoxide (CO)	20 to 2000 ppm

Furthermore, the robot is equipped with a temperature and humidity level detector. For this purpose, a DHT11 sensor is used, which contains a calibrated digital signal output. It has a capacitive humidity sensor and a thermistor for measuring temperature. The sensor works by sending out a start signal and then waiting for a response from the sensor. The sensor measures the time it takes for the signal to return to calculate both temperature and humidity [13]. This sensor measures the ambient temperature and humidity in the robot's vicinity. The recorded data is transmitted to the ground station through the micro-controller and Wi-Fi module. This information is useful for mapping temperature and humidity gradients, identifying areas of concern, and making informed decisions based on environmental conditions.

The micro-controller is responsible for controlling the motors that drive the robot's movement. It is connected to three motor drivers, which serve as interfaces between the micro-controller and the motors. These motor drivers receive signals from the micro-controller and provide the necessary power and control to drive the motors. The robot is equipped with two motors, each connected to a motor driver [14]. This configuration allows for precise control over the robot's locomotion, enabling it to navigate different terrains and obstacles effectively (figure 12).

Powering the entire system is a lead-acid battery that supplies 12 volts at 4 ampere-hours. This battery provides the necessary energy to operate the micro-controller, sensors, modules, and motors. The capacity of the battery ensures an extended operating time for the robot, allowing it to perform its tasks autonomously for a considerable duration. To enable remote control of the robot, the micro-controller receives signals from a transmitter circuit through a transceiver. This circuit allows users to remotely control the robot's movements, functions, and operations. The micro-controller interprets these signals and sends appropriate commands to the motor drivers to execute the desired actions.

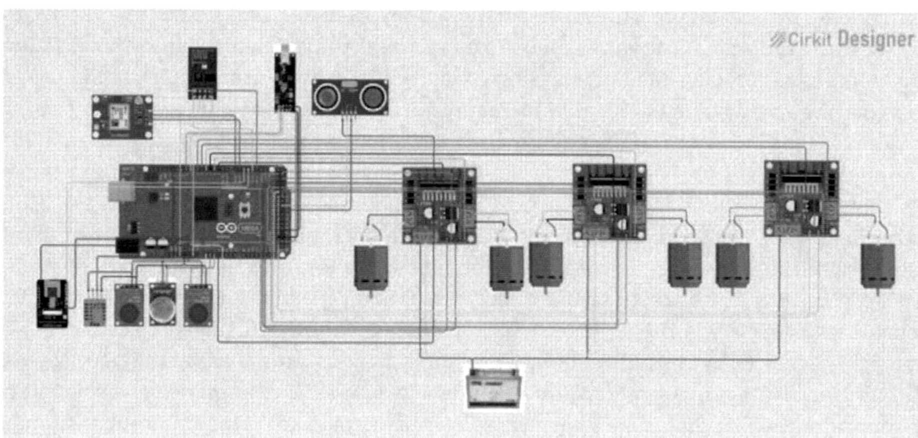

Figure 12. Circuit diagram of master circuit housed in the circuit box present on the robot.

All the sensors and modules connected to the micro-controller operate independently and continuously provide live data feeds. This includes the live camera feed, gas sensors, temperature sensor and ultrasonic sensor.

6. Results & testing

To finally create the robot, multiple rounds of testing, assembly and dismantling was required. Initially a prototype was created in the initial phase and then multiple additions were implemented (figure 13).

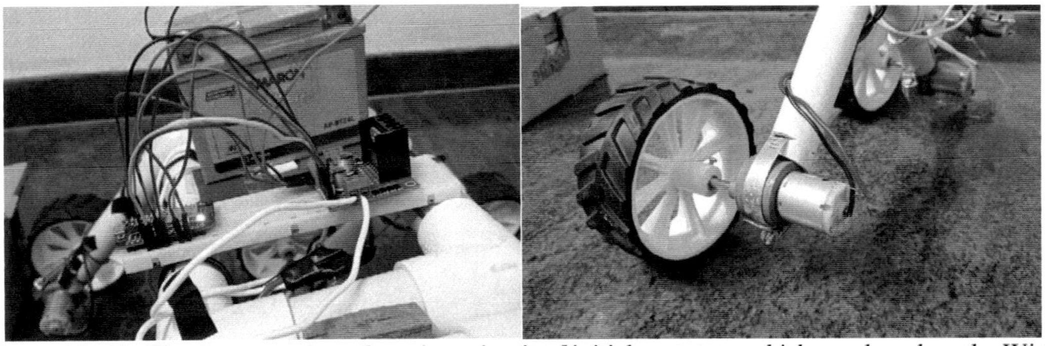

Figure 13. (a) Circuit connection of receiver circuit of initial prototype which was based on the Wi-Fi signals for movement (b) Motor and coupling connection.

In the initial prototype of the robot, some modifications were made to accommodate specific testing requirements. The battery was temporarily placed on the robot in the position originally designated for the accessory box. This adjustment was made to facilitate easy access and quick swapping of the battery during testing and development phases. Furthermore, instead of using radio frequency (RF) signals, the robot was controlled using Wi-Fi signals for testing purposes. To enable this functionality, an ESP-32 Node MCU micro-controller board was incorporated as the main control unit. The Node MCU board provides Wi-Fi capabilities, allowing for wireless communication and control (figure 14).

Figure 14. Connections of the microcontroller unit and the motor driver along with battery, the motor and wheel joint, and designed prototype respectively.

To control the prototype robot using Wi-Fi, an application was developed using the MIT App Inventor platform [15]. The application served as a user interface, providing control options and commands for the robot's movements and actions. When the user's phone connects to the Wi-Fi network created by the Node MCU, the application establishes a connection with the micro-controller, enabling remote control of the robot. Once the Wi-Fi connection is established between the user's phone and the Node MCU, the user can utilize the controls within the application to manipulate the robot's movements. By sending appropriate commands through the Wi-Fi network, the user can instruct the robot to move forward, backward, turn, or perform other specified actions (figure 15).

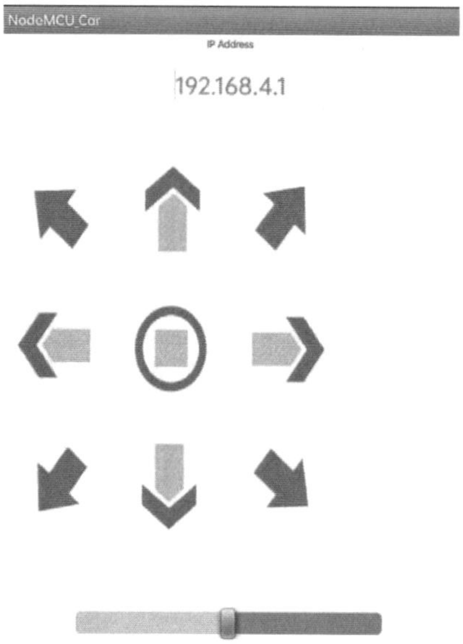

Figure 15. Screenshot of the user interface of the application created for the prototype.

In the next phase, the prototype (figure 16) was upgraded with high quality wheels, with better grip & greater traction abilities and climbing ability of the robot was improved upon and made more precise by redefining and improving upon the calculations and structure joints.

Figure 16. Prototype upgraded with better wheels and the motor updated with according of the fine-tuned calculations made.

The control of the robot was made more efficient by replacing Wi-Fi control with RF waves and designing a transmitter for the same (figure 17).

Figure 17. (a) Prototype robot upgraded with Radio Frequency Transceiver for receiving RF signals for control (b) Radio Frequency or Remote control (open format) created for the robot.

Following extensive testing of radio frequency control on the robot, the remote control was sealed, incorporating a switch for the battery-powered LED lights.

Figure 18. Upgraded prototype robot, climbing upon the hard 45° incline.

In the end, the prototype was upgraded with a host of sensors in the box and an accessory box to keep the payload as in when required.

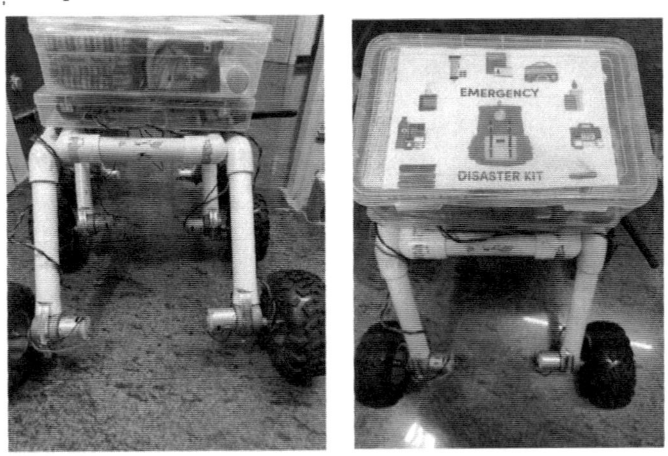

Figure 19. Upgraded prototype robot, with the two boxes – circuit box and accessory box stacked over each other.

A dashboard in the form of website was created to update all the sensor values and display the statistical analysis of the same in real time. the development of a website-based dashboard serves as a central hub for monitoring and managing sensor data effectively and efficiently. This dashboard represents an intuitive and accessible interface designed to provide users with a comprehensive view of the system's sensor readings and the analytical insights derived from them (figure 20).

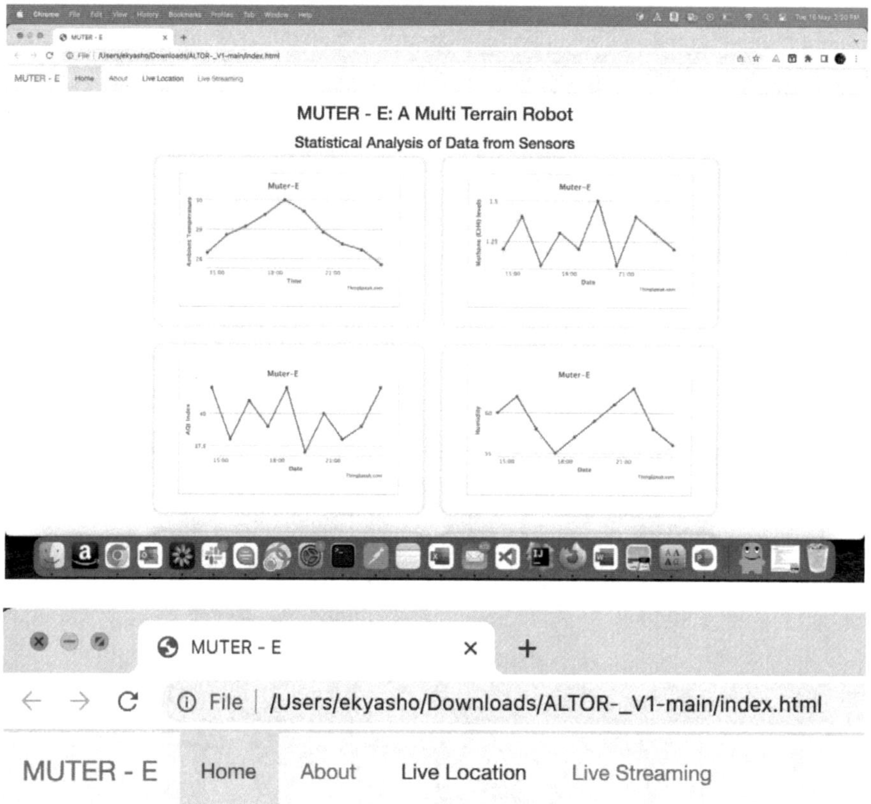

Figure 20. Multiple navigation tabs created in the website for the multi-terrain robot.

The website allowed the user to have the following features:

- Real-Time Data Updates: The website-based dashboard continuously receives and updates sensor data in real-time. This means that as new data becomes available from the sensors, it is instantly reflected on the dashboard, ensuring that users have access to the most current information.
- Multiple Sensor Integration: The dashboard is designed to aggregate data from various sensors deployed in the environment. This integration allows users to monitor multiple parameters simultaneously, such as temperature, humidity, air quality, CO levels, CH4 levels, and more.
- User-Friendly Interface: The dashboard features an easy-to-use and visually appealing interface. It may include graphical representations like charts, graphs, and tables that provide an at-a-glance overview of sensor readings. This visual representation simplifies data interpretation.
- Statistical Analysis: Beyond raw sensor readings, the dashboard offers statistical analysis tools. This can include calculations for averages, trends, historical data comparisons, and other analytical insights. These features enable users to gain a deeper understanding of environmental conditions and trends.
- Alerting and Notifications: To enhance usability, the dashboard can be equipped with alerting and notification mechanisms. If specific sensor values cross predefined thresholds or exhibit unusual patterns, the dashboard can trigger notifications to alert users or administrators, enabling prompt responses.

- Historical Data Storage: The website-based dashboard often includes the capability to store historical data for future reference, analysis, or reporting. This historical archive allows users to track changes over time and perform long-term assessments.
- Customization: Depending on user needs, the dashboard may offer customization options. Users can choose which sensors to display, set their own thresholds for alerts, and customize the appearance and layout of the dashboard to align with their preferences.
- Remote Access: One of the advantages of a website-based dashboard is the ability to access it remotely from various devices with an internet connection. This accessibility ensures that users can monitor and manage sensor data from anywhere, providing flexibility and convenience. The dashboard also includes the basic robot description & features.

7. Analysis

The multi-terrain robot based on the rocker bogie design, equipped with various sensors and connectivity features, is a versatile and advanced project. Here's a final analysis of its key components and functionalities:

- Rocker Bogie Design: The rocker bogie mechanism ensures excellent stability and manoeuvrability, enabling the robot to traverse challenging terrains with ease. The design created minimizes the risk of tipping over and ensures all wheels maintain contact with the ground, enhancing the robot's overall performance.
- Camera: The integrated camera serves as the robot's visual sensor, capturing real-time images or videos of its surroundings. The camera enables the robot to perceive its environment, recognize objects, and gather visual data for analysis and decision-making.
- GPS Module: The GPS module enables accurate positioning and navigation of the robot in outdoor environments. By leveraging signals from satellites, the robot can determine its precise location, create maps, plan optimal paths, and perform geolocation-based tasks effectively.
- Wi-Fi Module and Internet Connectivity: The Wi-Fi module facilitates wireless connectivity, allowing the robot to connect to the internet. This connectivity enables seamless data transmission and remote access to the robot's capabilities. By connecting to the internet, the robot can transmit sensor data and receive commands or instructions from a remote location, enabling remote monitoring, control, and data exchange.
- Ultrasonic Sensor: The ultrasonic sensor enhances the robot's obstacle detection capabilities. By emitting high-frequency sound waves and measuring their reflection times, the sensor can detect nearby obstacles. This information enables the robot to navigate complex environments by avoiding collisions and selecting safe paths.
- Gas Sensors: The inclusion of gas sensors expands the robot's capabilities to monitor environmental conditions. The gas sensors are designed to detect carbon monoxide, methane, smoke levels, and monitor the air quality index (AQI). These sensors provide crucial data for detecting potential hazards, such as gas leaks or poor air quality, allowing the robot to respond promptly and take appropriate actions.
- Data Display on Dashboard: The robot's connectivity to the internet and its data transmission capabilities enables the display of collected data on a dashboard in the form of a website. The dashboard is able to present real-time sensor readings, visualize environmental parameters, and provide insights into the robot's operations. Users can access the dashboard remotely and monitor the robot's sensor data, environmental conditions, and other relevant information in a user-friendly and intuitive manner.

After all these features integrated into the multi-terrain robot, it becomes a highly capable and versatile platform suitable for various applications. It can be deployed for tasks such as surveillance, environmental monitoring, disaster response, or exploration. The robot's ability to transmit data and display it on a dashboard enhances situational awareness and enables effective decision-making based on real-time information.

8. Applications

The multi-terrain robot project based on the rocker bogie design, equipped with various sensors and connectivity features, has a wide range of potential applications. Some of the key applications for this project include:

- Surveillance and Security: The robot's mobility and camera capabilities make it well-suited for surveillance and security applications. It can patrol areas, monitor perimeters, and detect any suspicious activities or intrusions. The real-time video feeds and sensor data can be transmitted to a central control station or security personnel, enabling prompt response and effective monitoring.
- Environmental Monitoring: The robot's gas sensors enable it to monitor various environmental parameters. It can detect gases such as carbon monoxide, methane, and smoke, providing early warnings for potential hazards like gas leaks or fires. The robot can be deployed in industrial settings, urban areas, or even homes to ensure air quality, detect pollution sources, and assess the impact of environmental factors.
- Search and Rescue Operations: In disaster scenarios, the robot can assist in search and rescue operations. Its ability to navigate through challenging terrains, such as rubble or debris, combined with its camera and sensors, allows it to locate survivors, identify hazards, and transmit vital information to rescue teams. This enhances the efficiency and safety of rescue missions.
- Exploration and Mapping: The multi-terrain robot is suitable for exploration tasks in remote or hazardous environments. It can traverse difficult terrains, capture high-resolution images, and create detailed maps. This capability is valuable for scientific research, geological surveys, or archaeological excavations, where the robot can gather valuable data and assist in the analysis of inaccessible areas.
- Industrial Inspections: The robot can be deployed for inspections in industrial settings, such as factories or oil refineries. Its mobility allows it to navigate complex layouts, while the camera and sensors enable it to visually inspect equipment, detect anomalies, or monitor environmental conditions. This aids in preventive maintenance, identifies potential risks, and reduces the need for human presence in hazardous environments.
- Agriculture and Farming: The robot can contribute to precision agriculture by monitoring crops, assessing soil conditions, and managing irrigation. Equipped with a camera and sensors, it can capture images, collect data on crop health, moisture levels, or nutrient content. This information helps optimize farming practices, reduce resource waste, and increase crop yields.
- Remote Monitoring and Telepresence: The robot's connectivity features make it suitable for remote monitoring and telepresence applications. It can be used as a remote presence device, enabling users to navigate and interact with the robot's surroundings from a remote location. This has applications in teleconferencing, remote inspections, or home monitoring, providing a sense of presence and control even when physically absent.
- Education and Research: The multi-terrain robot project serves as an educational and research platform. It allows students, researchers, and enthusiasts to gain practical experience in robotics, sensor integration, and data transmission. By exploring and customizing the robot's functionalities, users can conduct experiments, develop algorithms, and advance the field of robotics.

By leveraging its mobility, sensing capabilities, and connectivity features, the robot can address various challenges across industries, enhance safety and efficiency, and pave the way for innovative solutions in a wide range of domains.

9. Conclusions

In conclusion, the multi-terrain robot project based on the rocker bogie design, equipped with a camera, GPS module, Wi-Fi module, ultrasonic sensor, gas sensors, and internet connectivity, presents an advanced and versatile robotic platform. This project successfully combines mobility, perception, and data transmission capabilities to tackle a wide range of tasks in various environments. The rocker bogie

design ensures stability and mobility, enabling the robot to navigate challenging terrains while maintaining balance and preventing tipping. The integrated camera enhances the robot's perception abilities, allowing it to capture visual data for tasks like object recognition, navigation, and remote monitoring.
The GPS module provides accurate positioning and navigation, enabling the robot to operate in outdoor environments and perform tasks that require precise location information. The Wi-Fi module and internet connectivity allow for remote access and real-time data transmission, enabling remote control, monitoring, and data exchange with the robot. The ultrasonic sensor enhances obstacle detection, ensuring safe traversal by detecting objects in the robot's path and enabling collision avoidance. The gas sensors add environmental monitoring capabilities, enabling the detection of carbon monoxide, methane, smoke levels, and monitoring of the air quality index. This feature is essential for applications that require monitoring hazardous gases or assessing air quality. The ability to display data on a dashboard in the form of a website further enhances the project's usefulness. Users can remotely access the robot's sensor readings, environmental data, and operational information in a user-friendly manner, improving situational awareness and facilitating informed decision-making.
Overall, this multi-terrain robot project showcases a well-integrated and advanced robotic system that combines essential components and functionalities. It has the potential to be deployed in various applications, including surveillance, environmental monitoring, disaster response, and exploration tasks. The project's success highlights the importance of incorporating mobility, perception, connectivity, and data visualization features in creating effective and versatile robotic systems.

10. References

[1] Gupta A K and Gupta V K 2013 Design and development of six-wheeled Multi-Terrain Robot *CARE 2013 - 2013 IEEE International Conference on Control, Automation, Robotics and Embedded Systems, Proceedings* p 14095941

[2] Devi R, Dharrun B, Raj P G, Gowtham C and Kabilan A S 2021 Unmanned multipurpose all terrain rover using rocker bogie mechanism *Proceedings of the 6th International Conference on Communication and Electronics Systems, ICCES* p 21013109

[3] Das T, Sut D J, Gupta V, Gohain L, Kakoty P and Kakoty N M 2020 A Mobile Robot for Hazardous Gas Sensing *2020 International Conference on Computational Performance Evaluation, ComPE* p 19986505

[4] Buzzatto J and Liarokapis M 2023 The Omnirotor Platform: A Versatile, Multi-Modal, Coaxial, All-Terrain Vehicle *IEEE Access* **11** 27928

[5] Panneerselvam S, Bhoopesh D, Akhileshwaran K, Charan B and Aravindhan N 2021 Design and Fabrication of Rocker-Bogie Pesti-Bot Using Rotary Mechanism *2021 7th International Conference on Advanced Computing and Communication Systems, ICACCS* p 20799834

[6] Li J, He J, Xing Y and Gao F 2022 Dimensional optimization of rocker-bogie suspension for planetary rover based on kinetostatics and terramechanics *Proc Inst Mech Eng C J Mech Eng Sci* **236** 1

[7] Muthiah M, Nirmal K and Sathiendran R 2015 Low cost radio frequency controlled robot for environmental cleaning *IEEE International Conference on Circuit, Power and Computing Technologies, ICCPCT* p 15295573

[8] Rafiq A A, Yusuf M and Pujono 2018 Implementation of Digital Image Processing Using NI myRIO and Arduino Mega 2560 as Controller on Rover Bogie Robot *Proceedings - 2018 International Conference on Applied Science and Technology, iCAST* p 18796643

[9] Chauhan R, Bandopadhaya S, Dev M, Mishra A, Giri L I, Kharel B and Samal S R 2021 Design of Robotic Snake with ESP 32 CAM and Arduino *Proceedings - 2021 19th OITS International Conference on Information Technology, OCIT* p 21740248

[10] Nayanasitachowdary K and Padmaja M 2021 A Real and Accurate GPS based Environmental Monitoring Robotic System using IoT *Proceedings of the 5th International Conference on I-SMAC (IoT in Social, Mobile, Analytics and Cloud), I-SMAC* p 21527299

[11] Sen Gupta G, Mukhopadhyay S C and French J R 2008 Wireless communications and control module of a web-enabled robot for distributed sensing applications *Conference Record - IEEE Instrumentation and Measurement Technology Conference* p 10059234

[12] Prabhakaran S and Mathan N 2021 Safety Robot for Flammable Gas and Fire Detection using Multisensor Technology *Proceedings - 2nd International Conference on Smart Electronics and Communication, ICOSEC* p 21464026

[13] Tian W J and Geng Y 2009 A new household security robot system based on wireless sensor network *2009 2nd International Conference on Future Information Technology and Management Engineering, FITME* p 11084723

[14] Adnan M F F, Ahmad K A, Boudville R, Hussain Z and Husin N I 2020 Modelling and Simulation of Omnidirectional Mobile Robot with Line Follower and Obstacles Avoidance *Proceedings - 10th IEEE International Conference on Control System, Computing and Engineering, ICCSCE* p 20005155

[15] Munasinghe T, Patton E W and Seneviratne O 2019 IoT Application Development Using MIT App Inventor to Collect and Analyze Sensor Data *Proceedings - 2019 IEEE International Conference on Big Data, Big Data* p 19393174

IOP Conf. Series: Materials Science and Engineering 1295 (2023) 012006 doi:10.1088/1757-899X/1295/1/012006

Dynamic Modelling and Current Control for Switched Reluctance Motor Drives in Electric Vehicles

Muhammed A Ibrahim[1,2*], and Ahmed Nasser B Alsammak[1]

[1] Department of Electrical Engineering, College of Engineering University of Mosul, Mosul, Iraq
[2] Department of System and Control Engineering, Department, College of Electronics Engineering, Ninevah University, Mosul, Iraq

muhammed.ibrahim@uoninevah.edu.iq

Abstract. The inherent features of the Switched Reluctance Motor (SRM) make it a promising applicant for further exploration and examination in various industrial sectors, to mention but a few electric vehicles. The mechanical development of this motor is characterized by its simplicity and cost-effectiveness. Including current controllers within an SRM drive control system is of utmost importance due to their substantial influence on drive performance, dynamic response, and noise and vibration performance. This study aims to improve the SRM drive system to achieve reliability and efficiency. A complete SRM drive system is designed, constructed, and evaluated to accomplish this objective. The contribution of this study is that the consequence of the current control band of hysteresis current control (HCC) on the phase current has been intensely investigated on high-frequency torque ripple, average torque and average switching frequency. This investigation has been performed simultaneously with soft-hysteresis current control and hard-hysteresis current control. Simulation results proved that the soft HCC gives the best results. A trade-off between switching frequency and high-frequency torque ripple has been performed based on choosing the best current control of the proposed model.

1. Introduction

The global energy sector has faced the most difficult challenges and unclear futures in over half a century. British Petroleum's (henceforth BP) statistical review of World Energy 2022 highlights the expanding shortages and increasing prices. This emphasizes the continued relevance of energy 'security' and 'affordability' in addition to 'reduced carbon' when addressing the energy trilemma [1]. The rising demand for clean and abundant energy sources has guided growing interest in transportation electrification. Recently, a broad interest in (EVs) and their development has begun due to oil depletion, price rise, and the air pollution produced by cars with internal combustion engines (ICE). Therefore, the researchers are motivated to seek alternative energy sources to propel the car [2,3].

EVs will be the cornerstone for the future with lower carbon emissions. However, it would be advisable to allocate equal attention towards enhancing automobile efficiency and transitioning from coal to alternative renewable sources for electricity production. These two factors alone have the potential to bring about reductions in carbon emissions that are many times greater than those that will be brought about by the proliferation of electric vehicles over the next 20 years. Of course, each of these developments would coincide in a perfect world. However, decisions should be made in the real world, where resources are scarce. Figure 1 illustrates BP Energy Outlook's forecast for the rise of EVs from 2015 to 2035 [4].

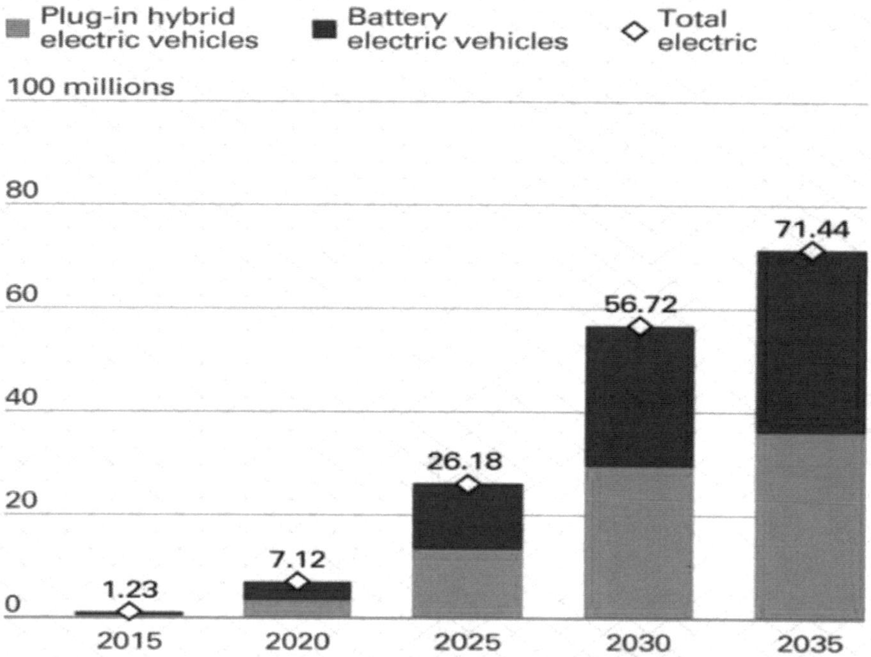

Figure 1. EV growth is predicted by BP energy outlook from 2015 to 2035 [4].

The phase current primarily determines the torque output in an SRM. The determination of the amplitude and timing of phase current in a specific phase is contingent upon the demand torque exerted by the mechanical load during a typical operational scenario. The torque is subsequently transformed into a reference current value that the controller utilizes. The phase current can be controlled at lower velocities so that it is at or close to the rated value. This will give the motor its maximum rated torque [5]. Due to the straightforward design of the stator, which incorporates a concentrated winding, and due to the absence of magnets and windings in the rotor, Switched Reluctance Motors (SRMs) exhibit notable suitability for challenging environmental conditions [6] and high-speed operational scenarios [7]. A doubly salient structure in the machine results in highly non-linear characteristics, which poses a significant challenge in control and modelling. The control of SRMs has been made easier due to recent advancements in power electronics technology [8,9]. Numerous academic researchers have endeavoured to minimize the losses associated with switching. The hysteresis/chopping current control technique is employed in SRMs to regulate the currents of the motor's phases within a specified range around a predetermined reference value [10]. The hysteresis control strategy is advantageous for operating SRMs over a wide speed range due to its ability to easily achieve the desired current. The control strategy involves activating the converter switches when the phase current falls below a predetermined lower threshold and deactivating these switches when the current exceeds a predetermined upper threshold [11,12].

Modern industrial applications in high-speed technology increasingly prioritize utilizing SRMs. The researcher is engaged in the development of environmentally sustainable systems. Electric motors play a pivotal role in facilitating energy conversion processes. This paper presents various control methods for SRMs. The paper introduces the machine to which various control methods are applied. The theoretical overview is verified with an accurate model implementation based on finite element analysis, and the potential to improve both the high-frequency torque ripple and acoustic noise characteristics of the SRMs is utilized based on the MATLAB simulation platform.

2. Operation of 8/6 switched reluctance motor

Figure 2 displays the cross-sectional view of a four-phase 8/6 SRM, along with the arrangement of the converter. The manuscript addresses a 4-phase 8/6 SRM configuration consisting of a rotor with 6 salient poles and a stator with 8 salient poles. The stator windings are arranged in a clustered configuration, where they are distributed into four sets that are equally spaced along the circumference of the motor. These sets are connected in series to form the motor phases. The rotor is designed without magnets and winding, enabling its construction with reduced inertia. The stator windings receive a separate power supply from four arms of an asymmetric half-bridge (AHB) power converter. Each converter phase consists of two transistors and two diodes that facilitate energy recovery on the DC source side. The operation of an SRM is elucidated by its magnetization characteristic, which represents the relationship between the machine's flux linkage (λ) and the rotor angular position (θ) as well as the stator current (I) [13,14].

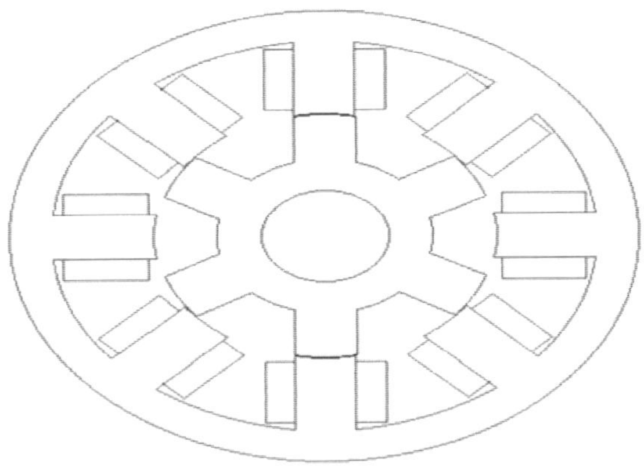

(a)

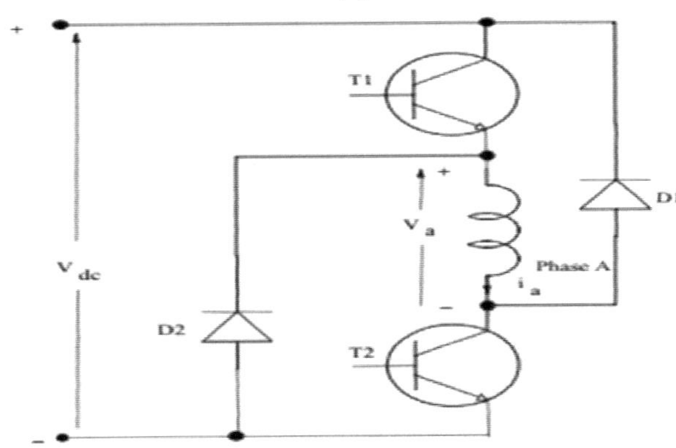

(b)

Figure 2. (a) Cross-section inspection of 8/6 SRM (b) Per phase AHB power converter.

The independent evaluation of the operational stages of the converter can be achieved by neglecting mutual coupling between phases, as represented in figure 3. The first operation stage occurs when both switches are closed, and the diodes are blocked, as presented in figure 3 (a). DC-link voltage is supplied to the phase, causing the current to rise while the switches remain turned on, which characterizes the magnetization process. When the switches are turned off, the energy accumulated in the winding affects the diodes to be forward-biased, offering a new path for the current to flow. This initiates the second stage of operation, as displayed in figure 3 (b). Negative DC-link voltage is supplied to the phase, decreasing phase current and de-magnetizing the winding. The diodes remain in conduction until the switches are turned on again or until the phase current reaches zero. A third stage can be used, where only one switch is kept on, and the other switch is turned off, as presented in figure 3 (c). Note that either the upper or lower switch can be used. This stage is known as the free-wheeling stage.

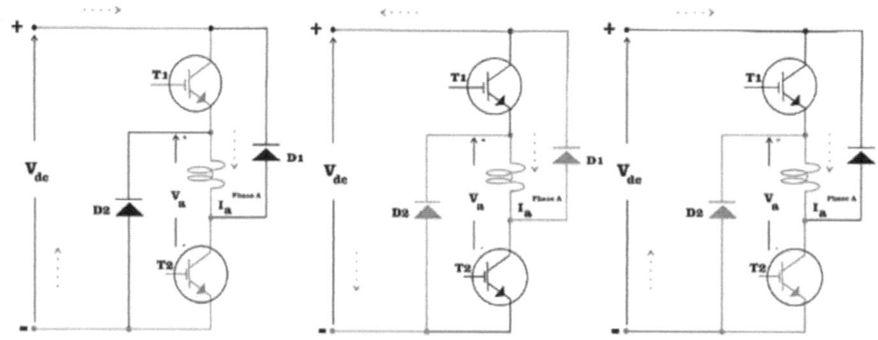

Figure 3. Operational stages of the AHB converter.

3. Modelling and simulation of the SRM

One of the critical elements in any control approach is the acquisition of an appropriate and accurate model that effectively characterizes the process under various operational conditions. Therefore. In order to accurately evaluate the performance of an SRM drive and the effectiveness of various control methods, it is imperative to have a reliable model of the SRM drive. The SRM model is required to accurately represent static and dynamic performance with sufficient precision [15,16]. The basic configuration for the SRM model is demonstrated in figure 4. The acquisition of magnetization and torque characteristics is critical to the model.

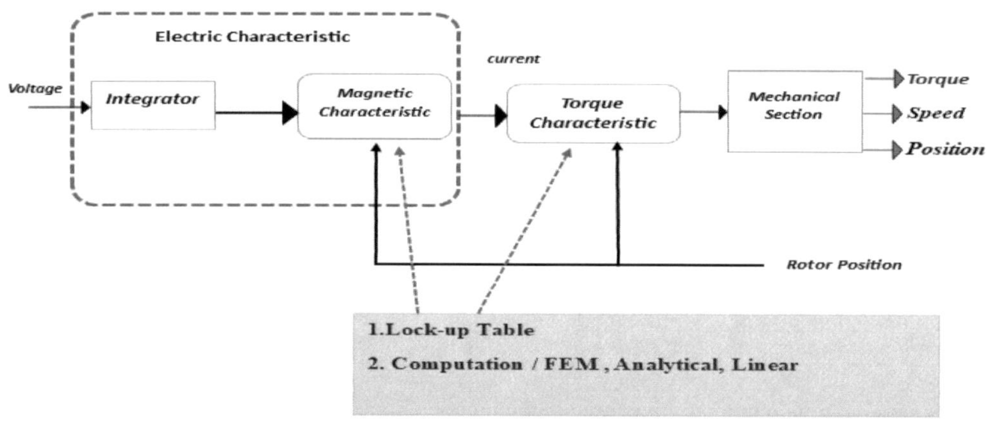

Figure 4. Block diagram of SRM model.

The movement equation represents the mechanical section of the 8/6 motor, which is derived from Newton's law. Similarly, the electric section of the motor is characterized by its voltage equations, which are based on Kirchhoff's law. In SRM, the voltage equation (1) is given by [16]:

$$v = R \times i + \frac{d\lambda}{dt} \tag{1}$$

Where R represents the resistance of the stator windings and λ represents the flux linkage. Magnetic saturation can be considered because the flux linkage is now a function of the current and the rotor position (θ). Therefore, the concept of flux linkage can be precisely characterized as equation (2):

$$\lambda(i,\theta) = L(i,\theta) \times i(t) \tag{2}$$

Where L is the machine inductance. Thus, the co-energy change rate to position can determine the electromagnetic torque value (Te) for a given current value [14]. The summation of the electromagnetic torques generated by the four phases results in the total electromagnetic torque exerted on the motor's rotor shaft. The equation of motion regulates the mechanical dynamics of both the motor and the load. When considering the operation of a motor, the dynamic equation for a general rotary machine can be formulated in the following manner [14,16]:

$$\frac{dw_m^{'}}{dt} = \frac{-B}{J} w_m + \frac{1}{J} T_m \tag{3}$$

The friction coefficient is denoted as B, while the rotor moment of inertia is represented by J. The concept of mechanical torque can be defined as:

$$T_m = T_e + T_L \tag{4}$$

The variable T_L represents the torque exerted on the machine's shaft.
The SRM utilized in this study is a classical 4-phase 8/6 machine. The non-linear properties of the machine are derived through a sequence of static two-dimensional finite element analyses based on the FEMM4.2 program [17,18]. After solving for machine parameters, the values are readily available in the results part of FEMM4.2 software. The flux lines in the investigated SRMs at two distinct positions (unaligned and aligned) can be viewed in figure 5 and figure 6, respectively. These positions' significance lies in giving this type of machine an overall non-linearity. Non-linear variation in inductance is caused by the fringe flux, which is more pronounced in the misaligned and intermediate positions.

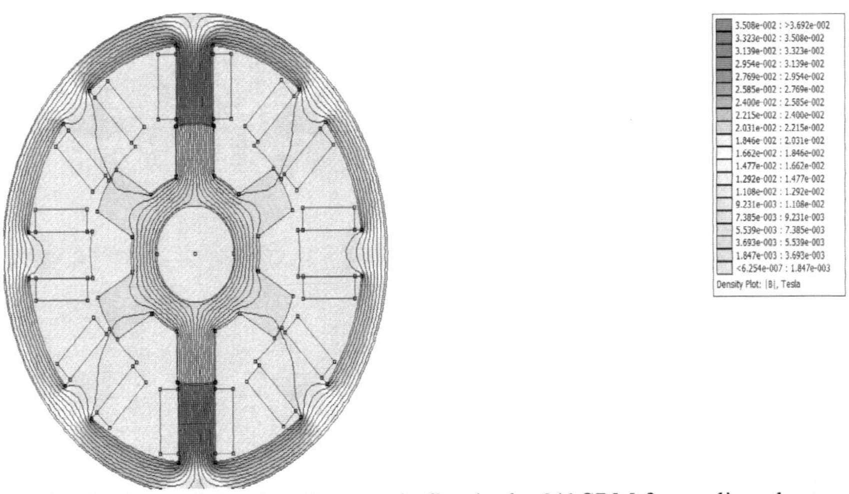

Figure 5. Distribution and density of magnetic flux in the 8/6 SRM for unaligned rotor position.

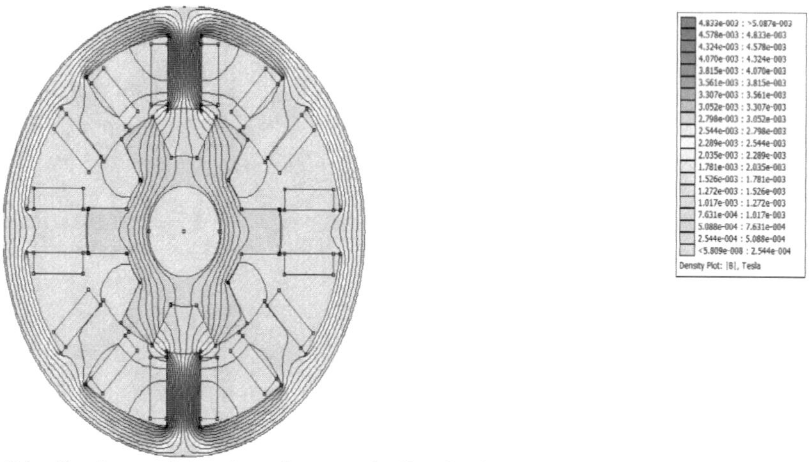

Figure 6. Distribution and density of magnetic flux in the 8/6 SRM for unaligned rotor position.

The obtained characteristics for flux-current-position are shown in figure 7, while the characteristics for torque-current-position are illustrated in figure 8.

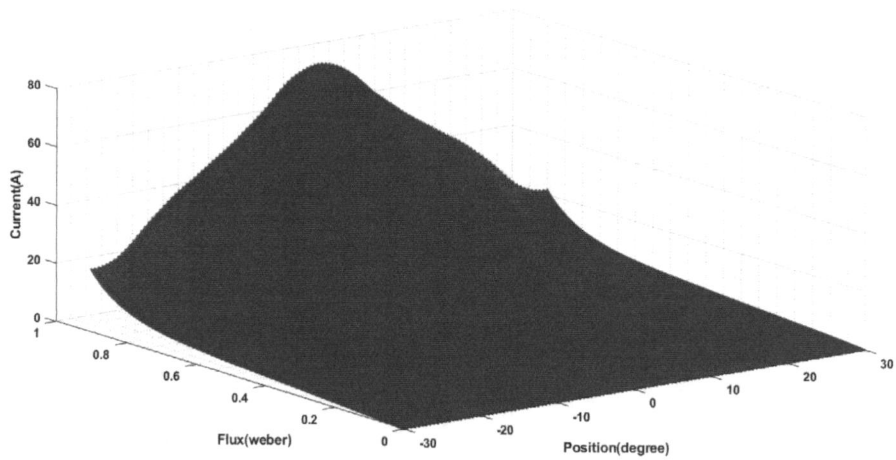

Figure 7. Flux – Current –Position characteristics (λ-i-θ characteristics).

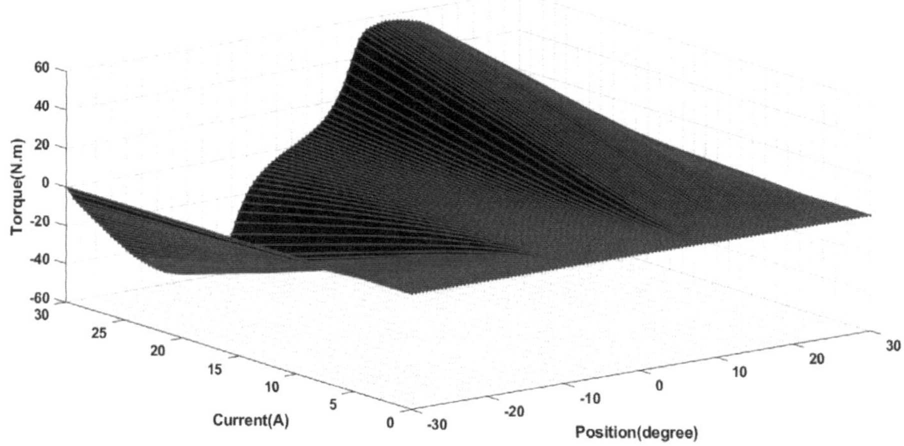

Figure 8. Torque – Current – Positions characteristics (T-i-θ characteristics).

With the obtained λ-i-θ characteristics and T-i-θ characteristics, the SRM complete model is implemented in MATLAB, utilizing look-up tables to mitigate the undesired approximations that result in imprecise outcomes and provide an inaccurate understanding of the machine. This model does not consider the mutual inductances due to their negligible value [19].

4. Current control approach

As earlier stated, current control is utilized for working below the base speed. Current control can be done in numerous techniques; however, due to the high non-linear behaviour of the SRM, hysteresis current control is frequently utilized. Hysteresis control needs the existence of an external control loop responsible to produce a reference current (I_ref). It is established on a torque or speed error. The block diagram can be investigated in figure 9. It displays that all phase currents are calculated and, dependent on the comparative phase position, are compared with the reference current so that the hysteresis development can be carried out. The hysteresis band is previously defined, and by comparing the current error with the limits of this band, the situations of AHB converter switches are controlled [16,18].

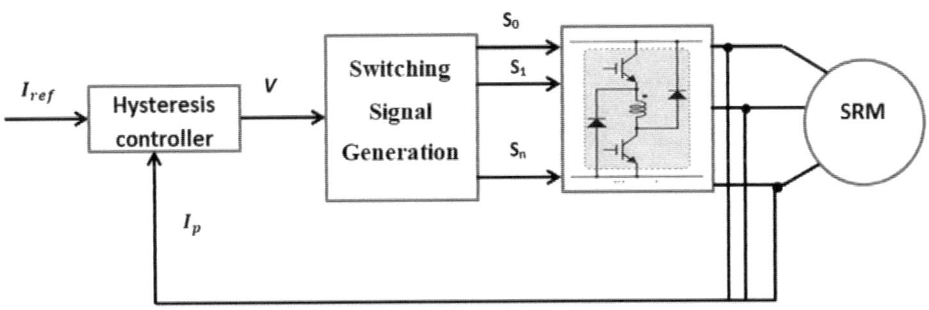

Figure 9. Simulink model of matrix converter.

5. Simulation and results

This partition respects the closed loop SRM drive system with the proposed hysteresis current controller of the intended SRM using MATLAB/SIMULINK software. The system response in numerous operating situations is investigated. Two different hysteresis current control (HCC) strategies (Soft HCC and Hard HCC) are implemented in the SRM drive system depending on how the switches of the AHB converter are activated. For the soft HCC strategy, the pulse width modulated (PWM) signal of the switch *T*1 is made ON and OFF, and the switch *T*2 is maintained ON during the dwell angle interval. For the hard HCC strategy, switches *T*1 and *T*2 are turned ON and OFF simultaneously. The choice of the control strategy used to activate the SRM phase impacts the machine's performance, as presented in figure 10, the current, voltage, flux, and torque waveform for both techniques.

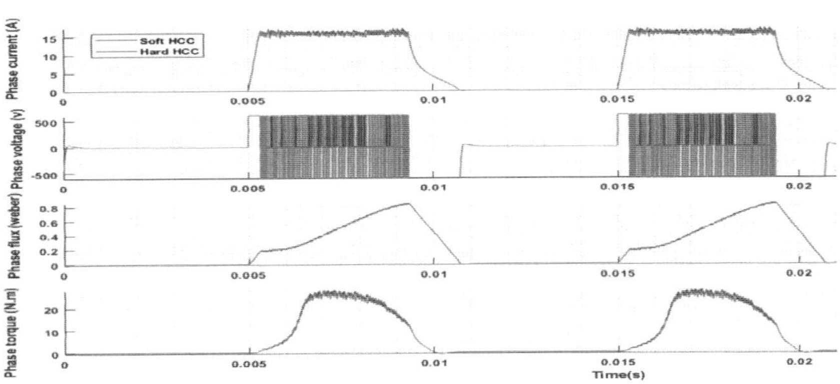

(a)

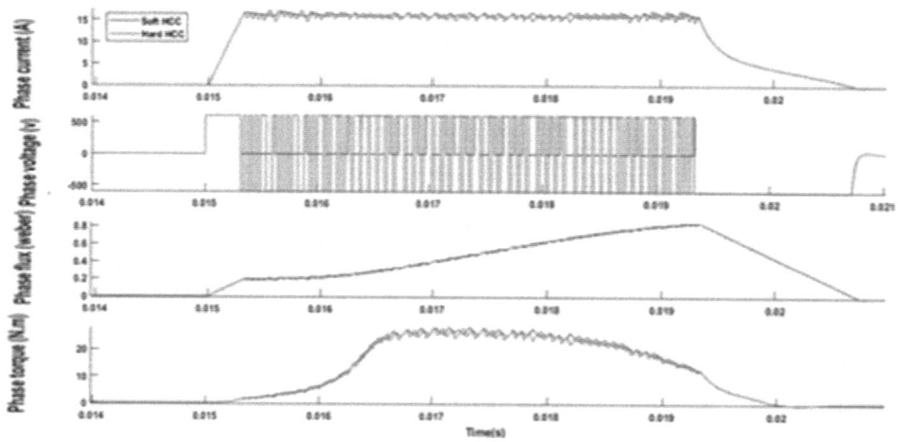

(b)

Figure 10. SRM drive response using soft HCC and hard HCC(a) current, voltage, flux, and torque waveform (b) zoom during one conduction period.

One can notice the impact of both strategies on the current wave. Consequently, the flux linkage and the torque during the conduction period. The wave shape of the flux also follows the current in terms of high-frequency ripples. The choice of the control strategy used for the SRM impacts the drive's performance.

The average switching frequency is determined by calculating the number of switching actions that occur through the alteration from the unaligned position (0°) to the aligned position (30°), the phase is excited during the initial half of the complete electrical period (0°-30°) and subsequently deactivated during the latter half (30°-60°) to prevent the generation of negative torque. Hence, the duration required for computing the mean frequency over the entire cycle should equal one electrical cycle [20].

The relationship between average switching frequency, speed, and torque can be comprehended and computed qualitatively utilizing a basic MATLAB script. However, precise outcomes can only be achieved by employing comprehensive simulation models. Figure 11 illustrates the relationship between switching frequency and speed for various current set-points while maintaining a constant hysteresis band of 1% in soft HCC.

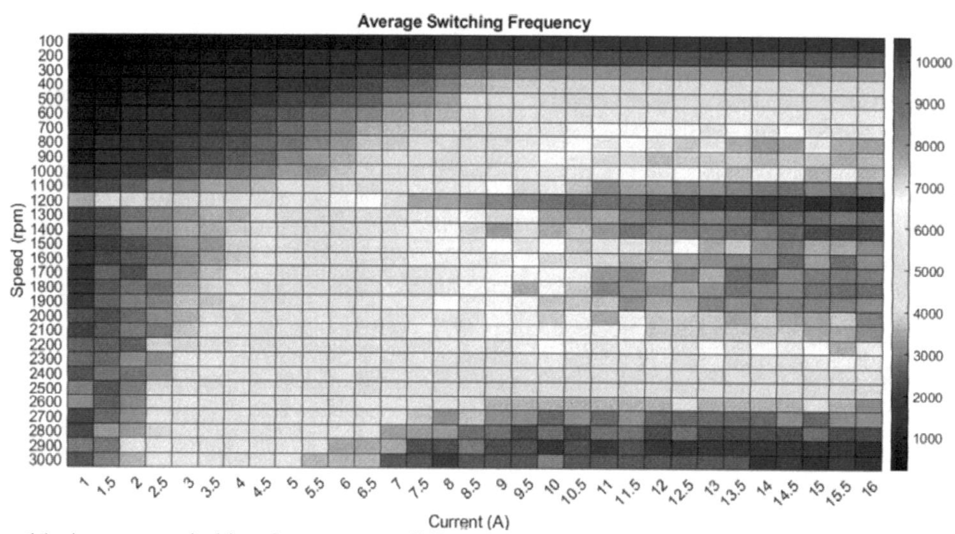

Figure 11. Average switching frequency at different operating points with 1%hysteresis band in soft HCC.

With a constant hysteresis band of 1% hard HCC, figure 12 shows the relationship between switching frequency and speed for various current set points.

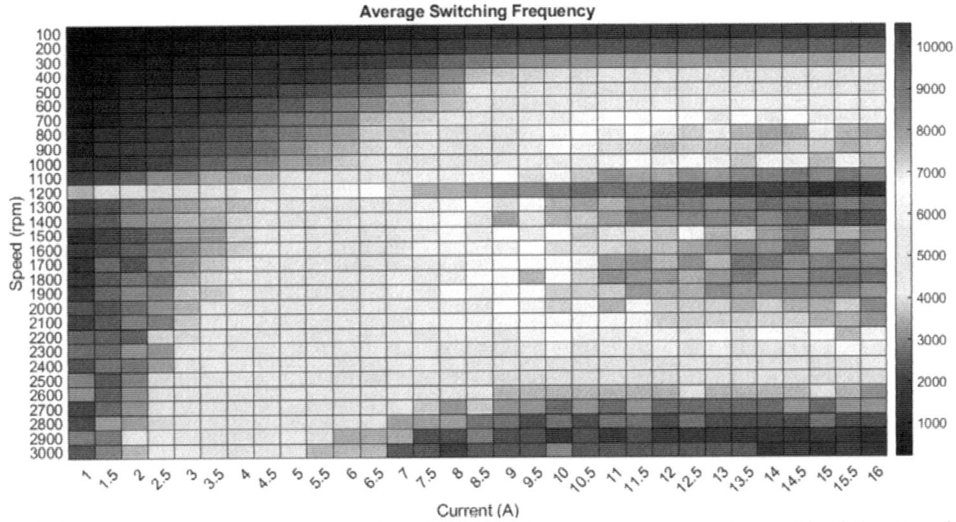

Figure 12. Average switching frequency function of different operating points with 1%hysteresis band in hard HCC.

Over the torque-speed range, it is demonstrated that the average switching frequency varies significantly. The findings can be used to calculate the switching losses of power converters. The high-frequency torque ripples are also directly proportional and influenced by the width of the current control band as presented in figure 13 for per phase current, voltage, and torque at 1000 rpm,16 A. while the turn-on angle and turn-off angle are kept constant (0,26) for hard HCC.

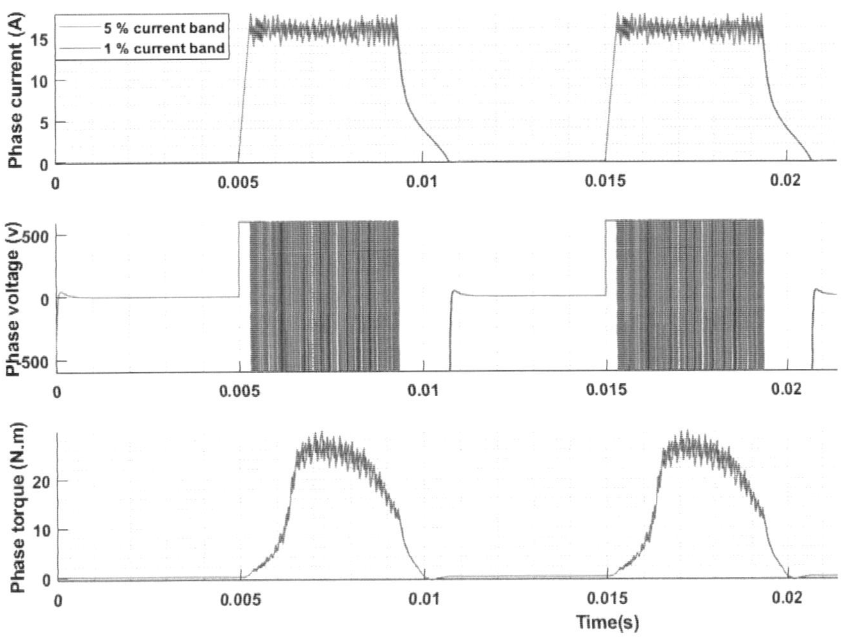

(a)

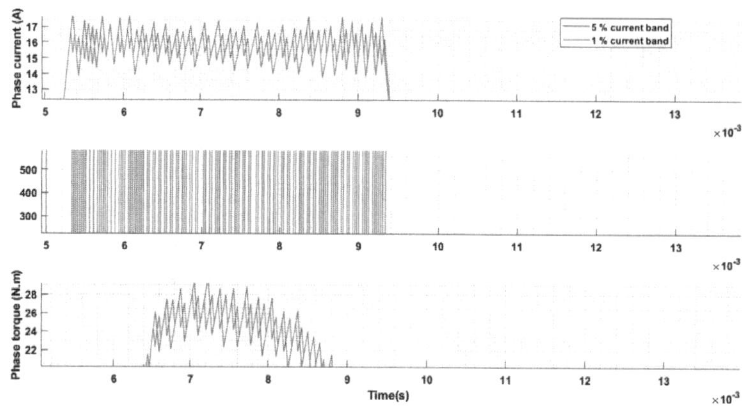

(b)

Figure 13. Variation of phase current, voltage, and torque for different hysteresis bands (hard HCC) (b) zoom during one conduction period.

Figure 14 displays the current, voltage, and torque per phase at a rotational speed of 1000 rpm and a current of 16 A in the context of the soft HCC system with the turn-on and turn-off angles being held constant (0°,26°).

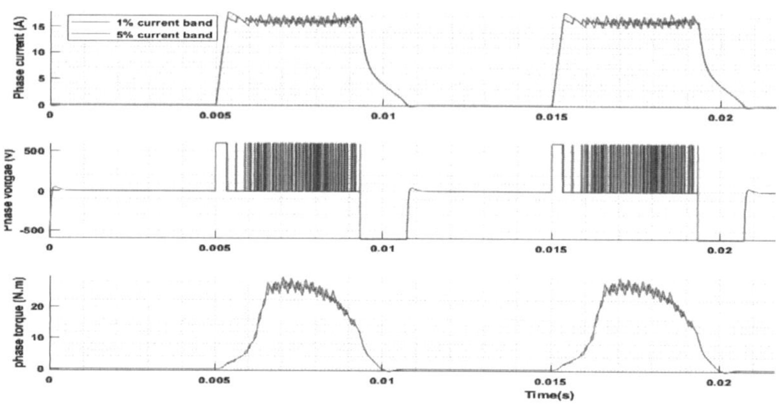

(a)

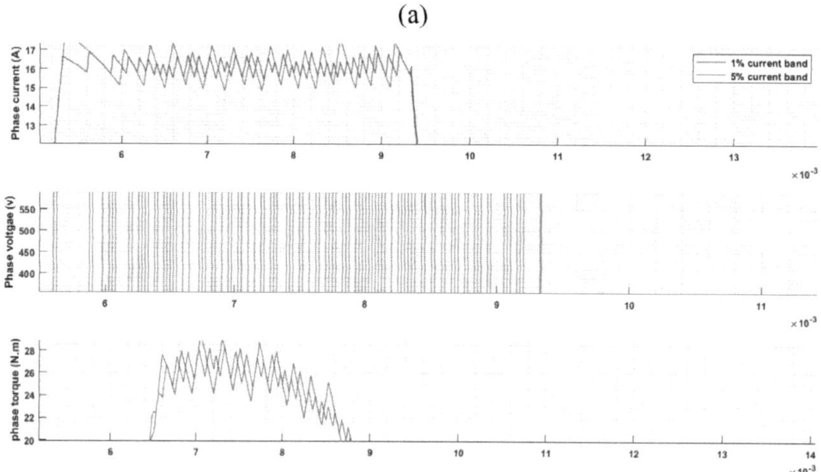

(b)

Figure 14. Variation of phase current, voltage, and torque for different hysteresis bands (soft HCC) (b) zoom during one conduction period.

A thorough understanding of switching frequency is crucial for designers, even during the initial stages of the design process. This knowledge is necessary for accurately calculating converter switching losses and conducting simulations and optimizations related to acoustic noise. It is crucial to acknowledge that the ideal amalgamation of switching frequency and hysteresis band is contingent upon many factors, including control strategy and operating conditions. In order to identify the optimal values for the switching frequency and hysteresis band, it is imperative to thoroughly evaluate the precise demands and compromises associated with a given SRM application.

6. Conclusion

The modelling of SRMs is highly intricate because of double saliency and extreme magnetic saturation. An accurate and reliable model is built for the employed 4 kW 8/6 SRM. The model can incorporate all non-linearities present in the machine accurately. The SRM can be easily modelled with the FEMM4.2 software due to internal calculations performed by this software compared to Ansoft Maxwell. In this study, an investigation is performed to examine the performance characteristics of the SRM drive in both hard and soft chopping modes, considering all drive system non-linearities. A proposed soft HCC is designed and implemented by enhancing the PWM pattern. It is found that soft chopping favours minimizing switching and reducing high-frequency torque ripple and current ripple at the lower speed. Due to its rapid response and model independence, hysteresis current control is highly advised for vehicle traction drives.

7. References

[1] Brandt A R, Millard-Ball A, Ganser M and Gorelick S M 2013 Peak oil demand: The role of fuel efficiency and alternative fuels in a global oil production decline *Environ Sci Technol* **47** 8031

[2] Needell Z A, McNerney J, Chang M T and Trancik J E 2016 Potential for widespread electrification of personal vehicle travel in the United States *Nat Energy* **1** 16112

[3] Mei J, Zuo Y, Lee C H T and Kirtley J L 2020 Modeling and Optimizing Method for Axial Flux Induction Motor of Electric Vehicles *IEEE Trans Veh Technol* **69** 12822

[4] Wadghule Tushar M 2013 A Review Of Plug-In Hybrid Electric Vehicles As New Storage And Transportation System For Smart Grid *International Journal of Engineering Research & Technology (IJERT)* **2** 1

[5] Blaabjerg F, Kjaer P C, Rasmussen P O and Cossar C 1999 Improved digital current control methods in switched reluctance motor drives *IEEE Trans Power Electron* **14** 563

[6] Dos Santos F L M, Anthonis J, Naclerio F, Gyselinck J J C, Van Der Auweraer H and Góes L C S 2014 Multiphysics NVH modeling: Simulation of a switched reluctance motor for an electric vehicle *IEEE Transactions on Industrial Electronics* **61** 469

[7] Xu Z, Li T, Zhang F, Zhang Y, Lee D H and Ahn J W 2022 A Review on Segmented Switched Reluctance Motors *Energies (Basel)* **15** 9212

[8] Banerjee R and Sensarma P 2023 Low-Cost Realization of Feature Position Estimation Scheme for Switched Reluctance Motor *IEEE Trans Power Electron* **38** 2850

[9] Vinh V Q and Ha V T 2023 Improved Torque Ripple of Switched Reluctance Motors using Sliding Mode Control for Electric Vehicles *Engineering, Technology and Applied Science Research* **13** 10140

[10] Elwakil E S and Darwish M K 2007 Critical review of converter topologies for switched reluctance motor drives *International Review of Electrical Engineering (I.R.E.E.)* **2** 50

[11] Srivastava K, Singh B K, Arya K V and Singh R K 2011 Simulation and Modeling of 8/6 Switched Reluctance Motor Using Digital Controller *International Journal of Electronics Engineering* **3** 241

[12] Sozer Y, Torrey D A and Mese E 2003 Automatic control of excitation parameters for switched-reluctance motor drives *IEEE Trans Power Electron* **18** 594

[13] Le-Huy H and Brunelle P 2005 A versatile nonlinear switched reluctance motor model in simulink using realistic and analytical magnetization characteristics *IECON Proceedings (Industrial Electronics Conference)* p 8923383

[14] Krishnan R 2017 *Switched reluctance motor drives: Modeling, simulation, analysis, design, and applications*

[15] Lin D, Zhou P, Stanton S and Cendes Z J 2009 An analytical circuit model of switched reluctance motors *IEEE Trans Magn* **45** 5368

[16] Fang G, Pinarello Scalcon F, Xiao D, Vieira R, Grundling H and Emadi A 2021 Advanced Control of Switched Reluctance Motors (SRMs): A Review on Current Regulation, Torque Control and Vibration Suppression *IEEE Open Journal of the Industrial Electronics Society* **2** 280

[17] Roy D and Sengupta M 2019 A novel and precise on-line digital measurement of unsaturated phase inductance of a switched reluctance motor *Sadhana - Academy Proceedings in Engineering Sciences* **44** 221

[18] Hamouda M and Számel L 2019 Accurate magnetic characterization based model development for switched reluctance machine *Periodica polytechnica Electrical engineering and computer science* **63** 3

[19] Ibrahim M A and Alsammak A N B 2022 Switched Reluctance Motor Drives Speed Control Using Optimized PID Controller *Przeglad Elektrotechniczny* **98** 11

[20] Fiedler J O, Fuengwarodsakul N H and De Doncker R W 2004 Calculation of switching frequency in current hysteresis controlled switched reluctance drives *PESC Record - IEEE Annual Power Electronics Specialists Conference* p 8190396

IOP Conf. Series: Materials Science and Engineering 1295 (2023) 012007 doi:10.1088/1757-899X/1295/1/012007

Maximum Torque per Ampere Technique for Switched Reluctance Motor Drive

Muhammed A Ibrahim[1,2*], and Ahmed Nasser B. Alsammak[1]

[1] Department of Electrical Engineering, College of Engineering University of Mosul, Mosul, Iraq
[2] Department of System and Control Engineering, College of Electronics Engineering, Ninevah University, Mosul, Iraq

muhammed.ibrahim@uoninevah.edu.iq

Abstract. Numerous researchers have investigated switched reluctance motors (SRMs) as a competitor to other electrical motors for usage in electric and hybrid vehicle appliances. SRMs are fault-tolerant and have a broad operating speed range. However, they have numerous drawbacks, such as high nonlinearity in machine models, complex control, and torque ripple. This study aims to improve the SRM drive system to attain maximum torque per ampere in a wide speed range. In the design and control of SRMs, maximizing torque per ampere (MTPA) is a crucial objective. MTPA refers to the strategy of achieving the maximum possible torque output for a given current input (amperes). This is essential for enhancing the efficiency and performance of the motor. The paper focuses on achieving MTPA in an SRM drive system-based commutation angle control. Utilizing a bacterial foraging optimization algorithm (BFOA), the suitable commutation angles at each operation point are determined. The intended technique is straightforward to realize and does not require complex algorithms. Simulation comparisons on an 8/6 SRM are also presented to validate the intended control technique. The results indicate that the MTPA has been enhanced by 6.961% compared to the conventional method.

1. Introduction

The switched reluctance motor (SRM) is a strong candidate for various variable-speed appliances due to its simple construction, durability, fault-tolerance capability, low manufacturing cost, and ability to operate at high speeds and temperatures. Decades of research have been conducted on it. However, compared to other traditional machines, one of its disadvantages is its higher electromagnetic torque ripple, which is initiated by its non-linear and double salient pole structure [1,2].

The enhancement of an SRMs performance can be achieved by striving for an optimal equilibrium between minimizing torque fluctuations and maximizing energy efficiency. Due to the inherent nonlinearity exhibited by the SRM, determining the optimal firing angles using analytical formulations becomes a challenging endeavour. Therefore, optimization algorithms play a crucial role in determining the optimal commutation angles [3,4]. To regulate the instantaneous torque of a SRM, torque-sharing functions, which can be either fixed or adjustable torque profiles, can be employed. The concept was initially proposed in reference [5] and subsequently enhanced in subsequent iterations [6,7]. The authors of reference [8] introduced a technique for implementing control rules, specifically maximum-torque-per-ampere control, and linear torque-current control. The efficacy of this approach was validated through both simulations and real-machine examinations. The MTPA

control approach is a widely recognized and extensively employed method for controlling both asynchronous and permanent magnet synchronous machines (PMSM) [9–11]. This approach is achieved by manipulating currents along the d and q axes through the utilization of specialized algorithms. The utilization of these techniques for the achievement of the MTPA approach is restricted by the SRM features. The control strategies employed for Switched Reluctance Motors (SRMs) differ significantly from those used for conventional electric motors. It is imperative to acknowledge that the set and potential control actions available during the execution of different methods of SRM working are contingent upon the range of drive speeds [12].
Currently, there is a growing interest in the utilization of SRM in contemporary high-speed industrial applications. There has been a shift in the focus of researchers towards the development of systems that are more environmentally sustainable. Electric motors play a essential role in the realm of energy conversion and serve as a pivotal enabling technology in the pursuit of environmental sustainability. This manuscript depicts the outcomes of a study on the modes of maximum torque per ampere (MTPA) in the low-speed zone of SRM operation. The outcomes pertaining to the advancement of control algorithms are utilized based on commutation angles optimization. The accurate representation of the SRM through modelling is a crucial element of this method. However, because of the nonlinearity exhibited by the SRM, an accurate simulation is developed using the FEMM4.2 software.

2. Operation of 8/6 switched reluctance motor

Salient poles are present in SRM on both the rotor and the stator. A concentrated coil is wound around each stator pole. Silicon steel laminations are used in the stator and rotor to reduce eddy currents. SRMs use different torque generation mechanisms based on variable reluctance. SRMs produce torque by magnetic reluctance variant, which has no magnetic excitation mechanism in the rotor and thus is highly robust. To reduce the reluctance in the magnetic circuit, the rotor moves toward the stator pole when the stator coils are energized, creating the stator magnetic field. Figure 1 depicts the torque generation mechanism [13,14].

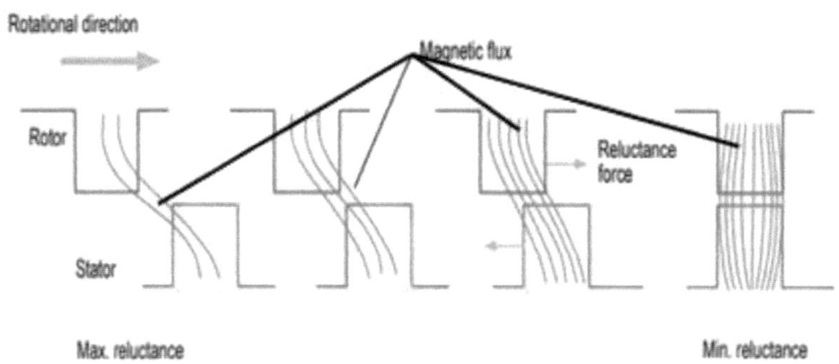

Figure 1. Torque production mechanisms.

The periodic variation characteristic of the machine inductance concerning the rotor position is known as the inductance profile. The profile period is a function of the number of rotor poles (Nr) of the machine, while the phase shift between phases is a function of the number of stator poles (Ns) and the number of phases(q). A theoretical inductance profile of the machine can be plotted as seen in figure 2; the coil inductance L varies with rotor position θ (Note that θ is obtained through an encoder attached to the machine shaft). The positive direction of rotation is in the counterclockwise direction. The coil is assumed to carry a constant current value. The generation of positive electromagnetic torque occurs exclusively when the inductance increases as the rotor moves closer to the aligned position[1,15].

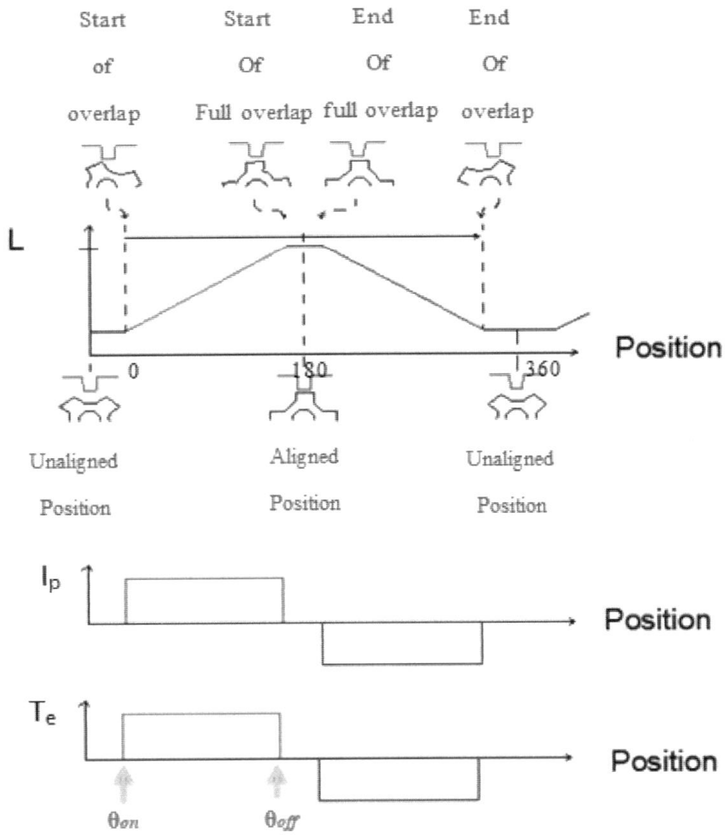

Figure 2. Variation of torque and inductance with rotor position.

The excitation opening at an angle θ_{on}, which is stated as the turn on angle θ_{on} and the angle when the de-magnetization is established is the turn off angle. Determining commutation angles (conduction angles) is contingent upon motor control. At the aligned position, the torque reverses direction. When the poles are separating, during intervals 180 through 360, the current must be cut off to remove the negative torque generation. The negative and positive torque impulses cancel when the machine rotates with a constant current flowing through the coil; thus, the average torque over an entire cycle is zero.

3. Modelling and simulation of the SRM

Acquiring a suitable and precise model that adequately characterizes the process under different operational conditions is a crucial component in any control approach. Consequently, to ensure a precise assessment of the operational efficiency of an SRM and the efficacy of different control techniques, having a reliable model of the SRM drive system is crucial. The utilization of the SRM model is necessary to effectively depict the static and dynamic performance while ensuring a satisfactory degree of accuracy [1,16].

The equation of motion defines the mechanical part of the 8/6 motor, which is derived from Newton's law. Similarly, the electric section of the motor is characterized by its voltage equations, which are based on Kirchhoff's law. In SRM, the voltage equation is given by [7]:

$$v = R \times i + \frac{d\lambda}{dt} \tag{1}$$

Where R represents the resistance of the stator windings and λ represents the flux linkage. Magnetic saturation can be considered because the flux linkage is now a function of the current and the rotor position (Θ). Therefore, the concept of flux linkage can be precisely characterized as:

$$\lambda(i,\theta) = L(i,\theta) \times i(t) \tag{2}$$

Where L is the machine inductance. Thus, the electromagnetic torque value (T_e) can be determined for a specified current value by the co-energy change rate to position [1].

$$T_e = \frac{dL(i,\theta)}{\Delta\theta} \times \frac{i^2}{2} \tag{3}$$

The total electromagnetic torque applied to the rotor shaft of the motor is obtained by summing up the electromagnetic torques generated by each of the four phases. The equation of motion governs the mechanical dynamics of both the motor and the load. When examining the functioning of a motor, the dynamic equation for a typical rotary machine can be formulated in a subsequent manner [1,4]. The study employed a classical 4-phase 8/6 machine for the SRM. The non-linear characteristics of the machine are determined by conducting a series of static 2D finite element analyses using the FEMM4.2 program [17,18]. Once the machine parameters have been solved, the corresponding values can be easily accessed in the results section of the Finite Element Method Magnetics (FEMM) software. The flux lines in the studied switched reluctance machines (SRMs) at two different locations, namely unaligned and aligned positions, are represented in figure 3 and figure 4, respectively. The importance of these positions resides in imparting a comprehensive non-linear characteristic to this machine. The presence of fringe flux in misaligned and intermediate positions is responsible for inducing non-linear variation in inductance.

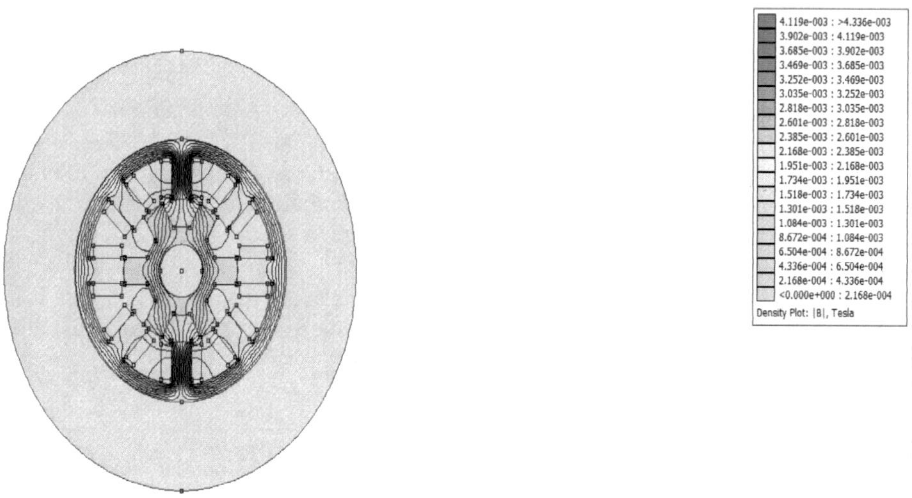

Figure 3. Distribution and density of magnetic flux in the 8/6 SRM for unaligned rotor position.

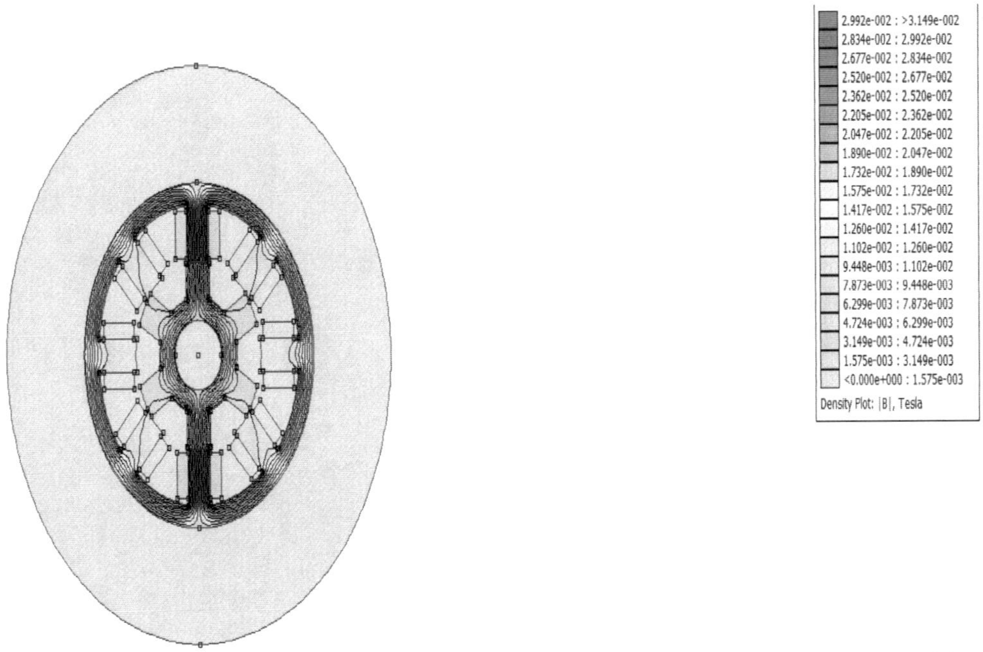

Figure 4. Distribution and density of magnetic flux in the 8/6 SRM for aligned rotor position.

The Flux – Current –Position characteristics and obtained is shown in figure 5 and Torque – Current – Positions characteristics is shown in figure 6.

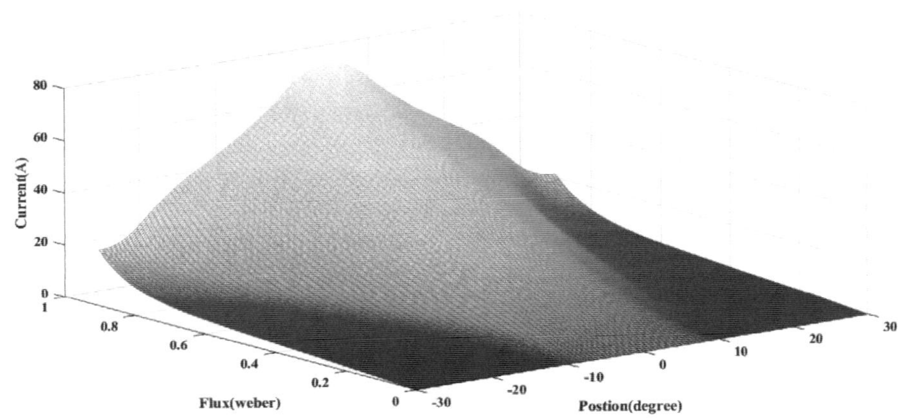

Figure 5. Distribution and density of magnetic flux in the 8/6 SRM for aligned rotor position.

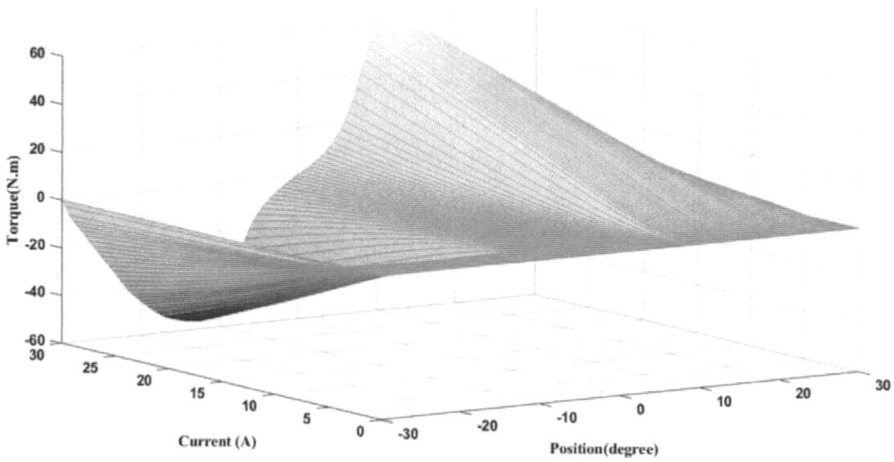

Figure 6. Torque – Current – Positions characteristics (T-i-θ characteristics).

The λ-i-θ characteristics and T-i-θ characteristics were used to implement the complete model of the SRM in MATLAB software. Look-up tables were employed to address the limitations of approximations, which can lead to imprecise results and an inaccurate understanding of the machine. The model does not consider the mutual inductances as they are deemed to have negligible value [19].

4. Optimizing the performance of an SRM by optimization techniques

The primary objective of optimizing the control variables of an SRM is to achieve optimal performance for the SRM drive system across various operating conditions. A proposed optimization methodology has been suggested, which utilizes a multidimensional approach and simulation techniques. The aim of this methodology is to improve the performance of the SRM (Switched Reluctance Motor) across a wide range of speeds. The optimization of commutation angles in a switched reluctance motor (SRM) entails determining the most appropriate commutation angles for the motor phases to attain desired performance attributes, including enhanced average torque, decreased current, and overall operational improvement. To optimize the conduction angles, it is essential to utilize a computational model capable of accurately calculating the dynamic torque shape of the machine using the specified commutation angles. The torque formulation in SRM drive is expressed as follows [20,21].

The maximum torque per ampere criteria performance index in an SRM drive system aims to optimize the motor for achieving maximum torque output per ampere of current see equation (9). This performance index focuses on improving the motor torque production efficiency, which can lead to higher overall performance and energy savings. The torque per ampere ratio measures the motor ability to generate torque efficiently for a given amount of current flowing through the motor windings. Maximizing the torque per ampere ratio allows the motor to produce higher torque output with reduced current consumption, improving efficiency and reducing power losses. This proposed control algorithm is designed to elucidate the fundamental principles of the maximum torque per ampere criteria methodology across a diverse range of speeds, utilizing the bacterial foraging optimization algorithm (BFOA). The commutation angles are methodically regulated within their respective ranges. To improve the torque per ampere (TPA) of the 8/6 SRM drive system. Figure 7 presents the formalization of the performance index selection process based on BFOA [22,23].

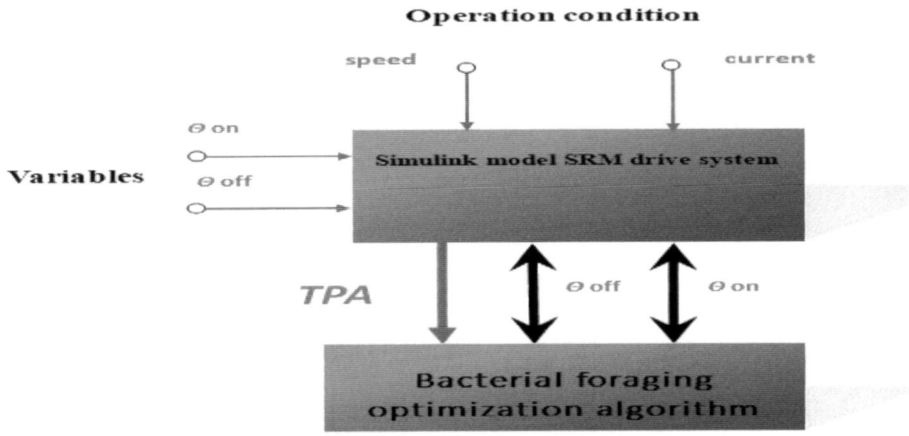

Figure 7. Description of the optimization problem.

5. Simulation and results

The primary objective of this study is to implement a partition that adheres to the proposed SRM drive system. More specifically, the aim is to optimize the commutation angles of the SRM under consideration. The MATLAB/SIMULINK software is employed for this objective. The investigation centres on analysing the system responses in diverse operating scenarios. The obtained simulation results are presented as a performance heatmap, which depicts the relationship between turn-on and turn-off angles. The performance curves can subsequently be integrated and employed as a precise design instrument to select the most suitable operating conditions in accordance with the criteria of the application. Figure 8 demonstrates the consequences of commutation angles on four parameters involving the RMS value of the phase current, average output torque, and torque per ampere at selected 1000 rpm 12 A single operating point.

The expressiveness of the results is enhanced when represented on three-dimensional charts as exemplified in figure 8 (A) the value of phase current (I_{prms}) is directly proportional to the commutation angle. Figure 8 (B) displays the average torque as θ_{on} and θ_{off} adjust. The maximum output average electromagnetic torque is not continually achievable due to the restraints on I_{prms} and the commutation angles. Figure 8 (C) shows the torque per ampere as θ_{on} and θ_{off} vary, which can be expressed to characterize the accurate qualities of SRMs. One commonly used measure to determine smart efficiency is the ratio of average output torque to the RMS value of phase current, T_{avg}/I_{prms},.

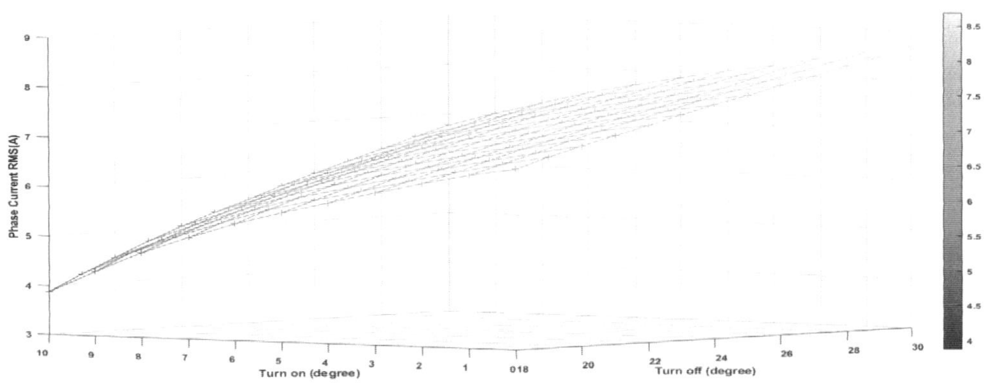

(a)

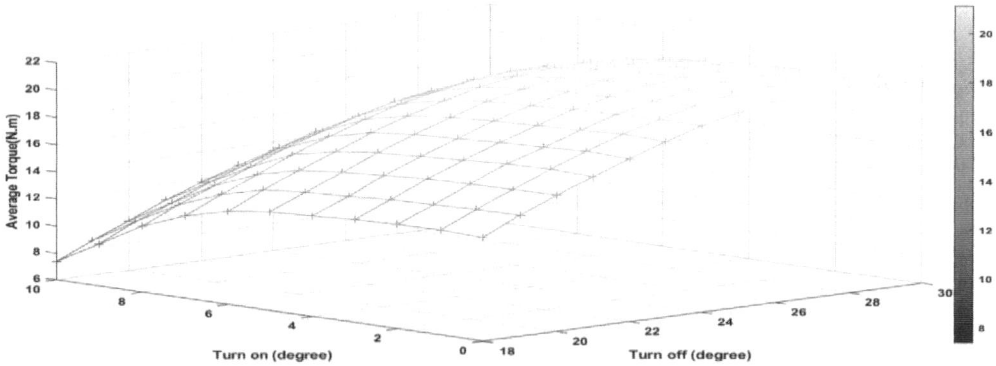

(b)

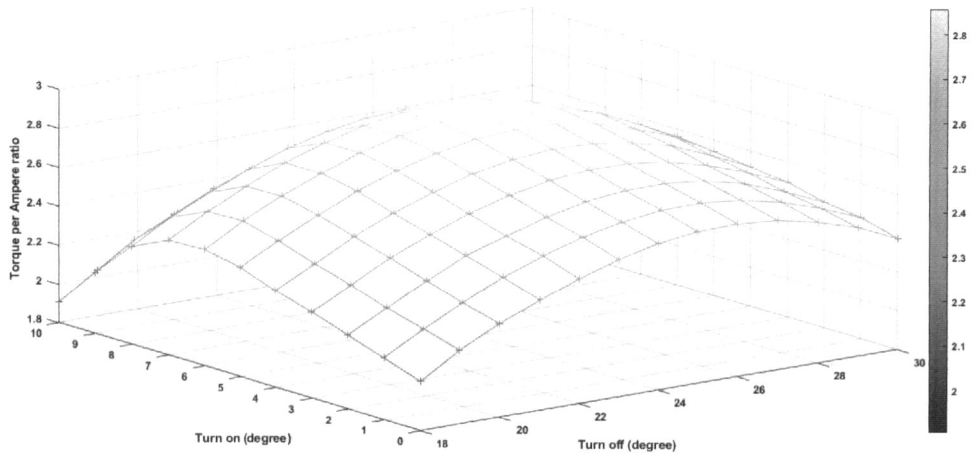

(c)

Figure 8. Three-dimensional charts of performance metrics for speed 1000 rpm I_ref=12 A : (a) RMS phase current, (b) average torque, (c) TPA.

The optimal performance index location varies in a single operating point. The commutation angles influence the torque quality and motor efficacy in the context of tuning. However, manually tuning these angles by simulation procedure for each operating point can be highly time-consuming. Effect of turn-on (θ_{on}) angle variation on TPA at a speed of 1000 rpm and I_{ref} 10 A is presented in table 1.

Table 1. Effect of turn-on θ_{on} angle variation on performance indexes at speed 1000 rpm and I_{ref} 10 A.

Commutation angles (degree)	T_{avg} (N-m)	I_{prms} (A)	TPA
$\theta_{on} = 0$ $\theta_{off} = 21$	12.87	6.08	2.116
$\theta_{on} = 2$ $\theta_{off} = 21$	12.69	5.77	2.199
$\theta_{on} = 4$ $\theta_{off} = 21$	12.6	5.45	2.311

$\theta_{on} = 6$	12.33	5.12	2.408
$\theta_{off} = 21$			
$\theta_{on} = 8$	11.01	4.67	2.357
$\theta_{off} = 21$			

Table 2 displays the effect of a turn-off (θ_{off}) angle variation on TPA at a speed of 1000 rpm and I_{ref}10 A.

Table 2. Effect of turn-off θ_{off} angle variation on TPA at a speed 1000 rpm and I_{ref} 10A.

Commutation angles (degree)	T_{avg} (N-m)	I_{prms} (A)	TPA
$\theta_{on} = 2$	11.9	5.61	2.121
$\theta_{off} = 20$			
$\theta_{on} = 2$	13.45	5.92	2.271
$\theta_{off} = 22$			
$\theta_{on} = 2$	14.81	6.21	2.384
$\theta_{off} = 24$			
$\theta_{on} = 2$	15.68	6.48	2.419
$\theta_{off} = 26$			
$\theta_{on} = 2$	15.96	6.74	2.367
$\theta_{off} = 28$			

The optimized commutation angle at a speed of 1000 rpm for different current values for maximum TPA is presented in table 3 by applying the proposed BFOA optimization algorithm presented in figure 7.

Table 3. Optimized commutation angles at a speed of 1000 rpm for maximum torque per ampere criteria.

Reference current	Commutation angles (degree)	Tavg (N-m)	Iprms (A)	TPA
16 A	$\theta_{on} = 5.809$	28.35	8.87	3.19
	$\theta_{off} = 24.57$			
14A	$\theta_{on} = 6.055$	23.86	7.81	3.05
	$\theta_{off} = 24.70$			
12A	$\theta_{on} = 6.481$	19.22	6.71	2.86
	$\theta_{off} = 27.75$			
10A	$\theta_{on} = 6.62$	14.85	5.71	2.6
	$\theta_{off} = 26$			
8A	$\theta_{on} = 7.054$	10.32	4.62	2.23
	$\theta_{off} = 25.81$			

The waveforms of phase "a" current, phase "a" voltage, phase "a" torque, and total electromagnetic torque for a rotational speed of 1000 rpm and a current of 10 A are illustrated in figure 9. The optimal TPA is 2.6 for selected operating points compared with a conventional approach for calculating commutation angles for TPA criteria.

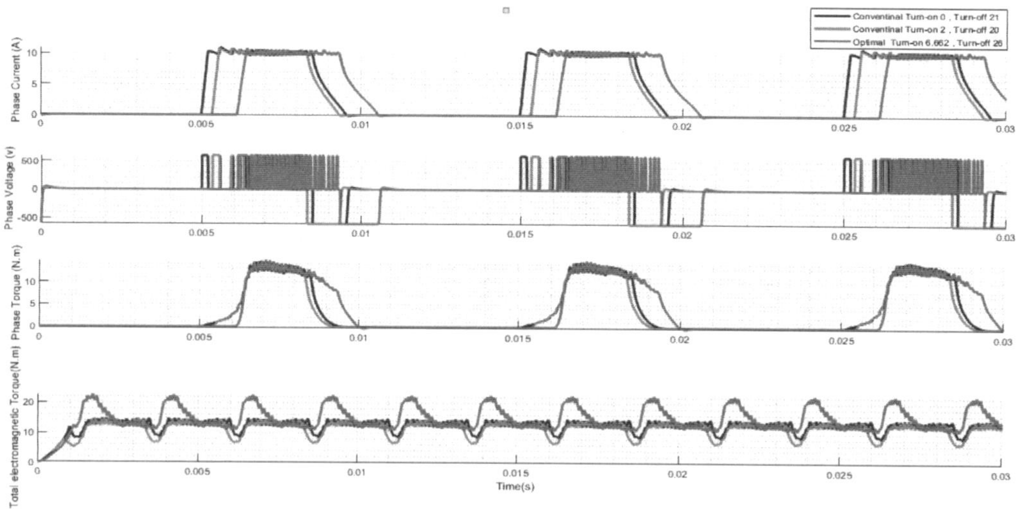

Figure 9. Waveform comparison based on optimized conduction angles at 1000 rpm, I_{ref} 10 A (a) Phase "a" current (b) Phase "a" voltage (c) Phase "a" torque (d)Total electromagnetic torque.

The obtainability of the consequences that result from the simulation platform has been observed. The results above are employed to evaluate the efficacy and dependability of the commutation angle control using bacterial foraging optimization algorithm (BFOA) for the SRM drive system. To examine the relationship between torque per ampere (TPA) and commutation angles, a simulation was conducted.

6. Conclusions

A BFOA control strategy based on commutation angles is presented to maximize torque per ampere overall operating points. A precise and dependable model is constructed for the 4 kW 8/6 SRM employed. The model is capable of precisely incorporating entirely nonlinearities present in the machine. Compared to Ansoft Maxwell, the SRM can be easily modelled with the FEMM4.2 software due to the internal calculations carried out by this program. The greatest control flexibility with the least added complexity is achieved by regulating the commutation angles separately at each operating condition. A novel approach utilizing optimization techniques has been introduced to enhance the control parameters of the SRM drive system, specifically targeting the improvement of TPA criteria. The findings indicate that the MTPA achieved a 6.961% improvement compared to the traditional approach.

7. References

[1] Krishnan R 2017 *Switched reluctance motor drives: Modeling, simulation, analysis, design, and applications*

[2] Lan Y, Benomar Y, Deepak K, Aksoz A, Baghdadi M El, Bostanci E and Hegazy O 2021 Switched reluctance motors and drive systems for electric vehicle powertrains: State of the art analysis and future trends *Energies (Basel)* **14** 2079

[3] Mademlis C and Kioskeridis I 2003 Performance optimization in switched reluctance motor drives with online commutation angle control *IEEE Transactions on Energy Conversion* **18** 448

[4] Fang G, Pinarello Scalcon F, Xiao D, Vieira R, Grundling H and Emadi A 2021 Advanced Control of Switched Reluctance Motors (SRMs): A Review on Current Regulation, Torque Control and Vibration Suppression *IEEE Open Journal of the Industrial Electronics Society* **2** 280
[5] Ilic'-Spong M, Marino R, Peresada S M and Taylor D G 1987 Feedback Linearizing Control of Switched Reluctance Motors *IEEE Trans Automat Contr* **32** 371
[6] Ishikawa H, Kamada Y and Naitoh H 2004 Instantaneous torque regulation for switched reluctance motors for the use in EVs *International Workshop on Advanced Motion Control, AMC* p 8057044
[7] Husain I and Ehsani M 1996 Torque ripple minimization in switched reluctance motor drives by PWM current control *IEEE Trans Power Electron* **11** 83
[8] Nakao N and Akatsu K 2016 Vector control specialized for switched reluctance motor drives *Electrical Engineering in Japan (English translation of Denki Gakkai Ronbunshi)* **194** 24
[9] Peng Z 2020 Analysis and implementation of constrained MTPA criterion for induction machine drives *IEEE Access* **8** 176445
[10] Libbos E, Ku B, Agrawal S, Tungare S, Banerjee A and Krein P T 2020 Loss Minimization and Maximum Torque-Per-Ampere Operation for Variable-Pole Induction Machines *IEEE Transactions on Transportation Electrification* **6** 1051
[11] Lin F J, Liu Y T and Yu W A 2018 Power Perturbation Based MTPA with an Online Tuning Speed Controller for an IPMSM Drive System *IEEE Transactions on Industrial Electronics* **65** 3677
[12] Barrass P G and Mecrow B C 1998 Flux and torque control of switched reluctance machines *IEE Proceedings: Electric Power Applications* **145** 34
[13] Ali Basher D and Ghayath Ibrahim O 2017 Propose Two Methods To Find Values of The Aligned And Unaligned Inductance of The Switched Reluctance Motor *Kirkuk University Journal-Scientific Studies* **12** 816
[14] Yu Q, Wang X, Cheng Y and Tian L 2019 *Analysis and Mathematical Models of Canned Electrical Machine Drives*
[15] Sozer Y, Torrey D A and Mese E 2003 Automatic control of excitation parameters for switched-reluctance motor drives *IEEE Trans Power Electron* **18** 594
[16] Lin D, Zhou P, Stanton S and Cendes Z J 2009 An analytical circuit model of switched reluctance motors *IEEE Trans Magn* **45 5368**
[17] Roy D and Sengupta M 2019 A novel and precise on-line digital measurement of unsaturated phase inductance of a switched reluctance motor *Sadhana - Academy Proceedings in Engineering Sciences* **44** 221
[18] Hamouda M and Számel L 2019 Accurate magnetic characterization based model development for switched reluctance machine *Periodica polytechnica Electrical engineering and computer science* **63** 3
[19] Ibrahim M A and Alsammak A N B 2022 Switched Reluctance Motor Drives Speed Control Using Optimized PID Controller *Przeglad Elektrotechniczny* **98** 11
[20] Sun X, Diao K, Lei G, Guo Y and Zhu J 2019 Study on segmented-rotor switched reluctance motors with different rotor pole numbers for bsg system of hybrid electric vehicles *IEEE Trans Veh Technol* **68** 5537
[21] Bizkevelci E, Ertan H B and Leblebicioglu K 2007 Effects of control parameters on SRM performance: A parametric search *International Aegean Conference on Electrical Machines and Power Electronics and Electromotion ACEMP'07 and Electromotion'07 Joint Conference* p 10014049
[22] Anon 2006 Biomimicry for Optimization, Control and Automation *IEEE Trans Automat Contr* **51** 1406
[23] Guo C, Tang H, Niu B and Boon Patrick Lee C 2021 A survey of bacterial foraging optimization *Neurocomputing* **452** 728

Smart Shopping Trolley with Automated Billing

K. Venkata Sai Prasad[1*], K. Karthik[1], O. Charan Kumar[1], Prathiksha B S[1], and Kavana Salimath[1]

[1] Department of Electrical and Electronics Engineering, Dayananda Sagar College of Engineering, Bengaluru, Karnataka, India

venkatasaiprasadkontam@gmail.com

Abstract. In today's context, people prefer buying groceries from supermarkets/hypermarkets. However, locating essential items in a supermarket can be time- consuming, and customers often have to wait in long billing queues to complete their purchases. The COVID-19 pandemic has further complicated matters, as maintaining social distance during the billing process is challenging. To address this significant challenge, this research proposes a smart trolley based on the Internet of Things (IoT) with an advanced billing system, aiming to make shopping easier, more secure, and to eliminate the need for standing in long queues. The proposed system comprises a smart trolley equipped with an LCD display, barcode scanner, and a Raspberry Pi. This innovative model aims to revolutionize the shopping experience by streamlining the process and mitigating administrative issues. The proposed framework holds promise for widespread implementation in real-world scenarios.

1. Introduction

Throughout history, humans have continuously innovated to meet their evolving needs, driven by the quest for greater independence and streamlined processes. However, shopping remains a time-consuming task, with traditional shopping baskets and carts causing long queues and overcrowding at payment counters in most shopping centers. To address these challenges and improve the shopping experience, we have introduced the Smart Shopping Cart. This innovative system features a functional smart trolley equipped with a Raspberry Pi, touchscreen LCD display, and Barcode Scanner. Customers can easily select their products, and the system automates the billing process for seamless transactions. The Smart Shopping Cart aims to revolutionize shopping, offering convenience and security while reducing congestion at checkout counters. By integrating technology into shopping carts, our goal is to create a positive impact on the overall shopping experience for all consumers. Firstly, the system's ability to automate the billing process based on scanned items leads to faster and more efficient checkouts. As customers shop and scan products, the system continuously calculates the total bill in real-time. This eliminates the need for manual item-by-item calculations at the checkout counter, resulting in shorter transaction times. The convenience of automated billing ensures that customers can complete their purchases swiftly, thereby reducing the length of queues.

Additionally, the integration of a QR code-based payment option adds to the efficiency of the system. Once the shopping is done, customers can quickly and securely make payments by scanning the generated QR code using a payment app or device. This eliminates the need for physical cash transactions or card swiping, further reducing the time spent in the payment process.

2. Literature review

In their study, [1] discussed the swift scanning of barcodes, with the integrated weight measurement conducted through a load cell module located at the cart's base. The weight data is continuously transmitted to the database, along with the corresponding product number sent to the server. This synchronization of the weight sensor with the system enables the identification of added or removed products, facilitating accurate billing.

In [2] authors developed a medical label barcode acquisition and identification system based on deep learning, and its feasibility was successfully verified. The system demonstrates high real-time performance and decoding rates, meeting the requirements for practical engineering applications. By combining neural network localization with traditional image processing, the method overcomes positioning challenges like distortion, fouling, and obstructions, which are difficult to address using conventional techniques. As a result, the software's decoding capability for linear distortion Data Matrix codes has been significantly improved.

In [3] authors explored a model for visual object recognition, thoroughly testing its performance in various complex and dynamic environments. The system exhibits object recognition capability even for objects of varying sizes. The recognized objects are converted into an audio format, complete with their location tags within the image and their corresponding accuracy. This intelligent vision system utilizes TensorFlow and neural network algorithms, aiming to minimize localization errors, implement dynamic thresholds, and enhance overall efficiency for superior performance.

In [4] authors implemented an innovative, cost-effective LiFi system in shopping malls to improve the shopping experience. Their system utilizes Li-Fi, RFID, and VLC technology, effectively resolving the existing challenges in the shopping system. By incorporating the latest communication technology, Li-Fi, into the shopping experience, the proposed model presents a promising solution for future generations and offers a safer, more efficient, and productive shopping environment.

In [5], author introduced a novel smartphone shopping application that utilizes barcode scanning and Beacon technology to provide customers with nearby store coupons. This app enables users to shop seamlessly both online and offline, benefiting from discounted prices and efficient management of their mobile shopping carts. By visually checking products and obtaining quick details, customers can make informed purchase decisions. This approach effectively attracts both showrooming and reverse showrooming, offering advantages to enterprises. However, it is acknowledged that the application is still a work in progress, necessitating further collaborations with online and offline stores to achieve its full potential. As the smartphone shopping application continues to evolve, continuous improvement efforts are required before its official release.

In [6], author presented an affordable and efficient Smart Shopping Cart that leverages IoT (Internet of Things) technology. This system is specifically designed for use in places like Walmart and supermarkets, aiming to reduce labour and enhance the shopping experience for customers. By automating the billing process, the system eliminates the need for customers to wait in long checkout queues. The proposed system accommodates two shopping approaches: predefined lists and random shopping. For the former, it provides the most optimal route to pick up items from different racks in Walmart. Additionally, the system allows Cart-to-Cart communication, enabling customers to share their shopping lists with co- shoppers and facilitating parallel shopping using multiple carts. These features not only save time but also simplify the shopping process. Furthermore, the system design includes theft detection capabilities, effectively combating shoplifting incidents.

3. Proposed system

The proposed system is an advanced automated shopping trolley system designed to revolutionize the traditional shopping experience (figure 1). It incorporates several cutting-edge technologies and devices to provide customers with a seamless and convenient shopping journey. At the core of the system is a barcode scanner that allows customers to easily self-scan the products they wish to purchase. By simply scanning the product's barcode, the system instantly retrieves essential information such as product name, price, and other details. This not only speeds up the shopping process but also ensures accuracy in product selection.

To facilitate the interaction between the customer and the system, a Raspberry Pi, acting as the central control unit, is integrated into each shopping trolley. The Raspberry Pi handles the processing of data and communication between the different components, ensuring smooth operation and reliable performance.

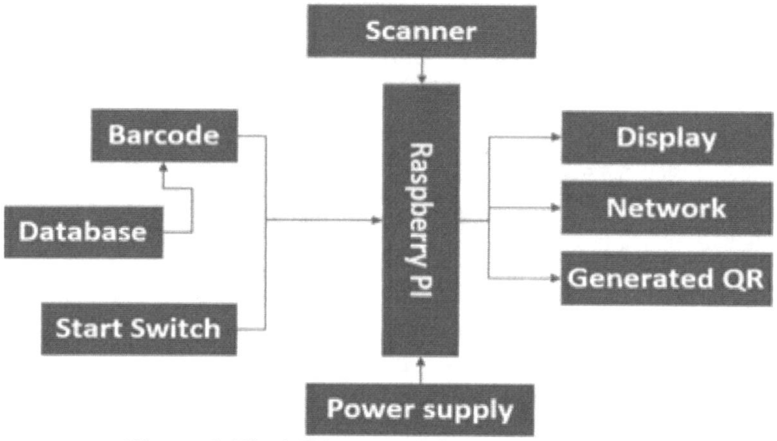

Figure 1. Block Diagram of the proposed model.

The system incorporates a user-friendly touch screen LCD display as the interface for customers. This display provides a graphical user interface (GUI) that allows customers to view and update product details. For example, if a customer wants to add or remove an item from their purchase list, they can easily do so through the touch screen interface. The display also provides real-time information, such as the running total of the bill, making it convenient for customers to keep track of their expenses.

To enable efficient data management, each shopping trolley is equipped with a wireless smart device. This device keeps a record of all the scanned products, creating a virtual shopping cart for each customer. It communicates wirelessly with the supermarket's backend database, which contains comprehensive information about the products in stock. By accessing this database, the system retrieves crucial details such as product ID, price, stock availability, expiry date, and even the purchase history of the specific item. One notable feature of the system is its ability to check the expiry date of scanned products. If a product is found to be expired or close to expiration, the system immediately notifies the customer through the touch screen display. This ensures that customers are aware of the product's condition and can make informed decisions. When it comes to billing, the system automates the process based on the scanned items. It calculates the total bill by summing up the prices of the scanned products. Once the customer has finished shopping, they can proceed to the payment stage. By selecting the payment option on the touch screen display, the system generates a QR code that represents the total bill. This QR code can be scanned by a payment app or device, enabling secure and hassle-free payment [7].

Additionally, the proposed system includes a feature for theft detection purposes. A security page is created specifically for the security personnel, providing them with a comprehensive overview of the recently bought products by customers and their names. This feature serves as an additional layer of security to cross-check the products at the checkout counter, effectively preventing theft.

The security page is accessible to the security personnel through a dedicated interface. It displays a list of all the products purchased by customers, along with their corresponding names. This enables the security personnel to compare the scanned items with the products in the customer's possession during the checkout process. If any discrepancies or suspicious activities are detected, the security personnel can take appropriate measures to address the situation promptly. By having this security page, the system enhances the overall security measures in the supermarket. It acts as a deterrent against theft by providing a mechanism to verify the legitimacy of the purchased items. This not only protects the supermarket's inventory but also ensures a safe and secure shopping environment for all customers.

The integration of the security page into the proposed system adds an important layer of theft prevention and enhances the overall effectiveness of the automated shopping trolley system [8].

IOP Conf. Series: Materials Science and Engineering 1295 (2023) 012008 doi:10.1088/1757-899X/1295/1/012008

4. Execution

The execution of automated shopping trolley system described in the paper involves a series of steps to ensure its successful deployment and operation. Initially, the hardware setup phase focuses on configuring the essential components within each shopping trolley (figure 2). This includes the integration and proper wiring of the barcode scanner, Raspberry Pi, and touchscreen LCD display. Careful attention is given to establishing seamless connectivity and communication between these devices to ensure their synchronized functioning. The subsequent step entails the development of the software system that will govern and manage the automated shopping trolley system. This encompasses the creation of an intuitive graphical user interface (GUI) specifically designed for the touchscreen LCD display (figure 2). The GUI empowers customers to effortlessly scan products and modify their details, enabling actions such as adding or removing items from their purchase list. Moreover, the software system incorporates sophisticated barcode scanning capabilities, allowing the system to accurately interpret and process scanned barcodes [9]. By leveraging this functionality, the system retrieves pertinent product information, including the product's name, price, and stock availability. Such data plays a pivotal role in ensuring precise billing and effective stock management. To establish real-time access to comprehensive product information, the software system is seamlessly integrated with the supermarket's backend database. This integration enables the wireless smart device within each shopping trolley to establish a connection with the database. By leveraging this connection, the system retrieves up-to-date product details, such as unique product identifiers, prices, expiry dates, and purchase history. This real-time data retrieval ensures the system operates with accuracy and efficiency, empowering customers with reliable information. To enhance security measures, a dedicated security page is developed exclusively for security personnel. This page displays a comprehensive list of recently purchased products and their corresponding names. This serves as a valuable tool for cross-checking items during the checkout process, effectively preventing theft and ensuring the authenticity of purchased items.

Thorough testing and optimization procedures are diligently conducted to validate the seamless operation of the system. This involves meticulous bug identification and resolution, as well as incorporating user feedback to further refine the system's performance.

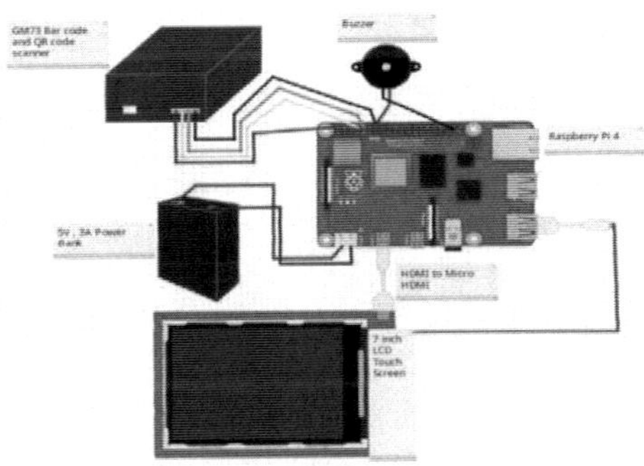

Figure 2. Hardware schematic.

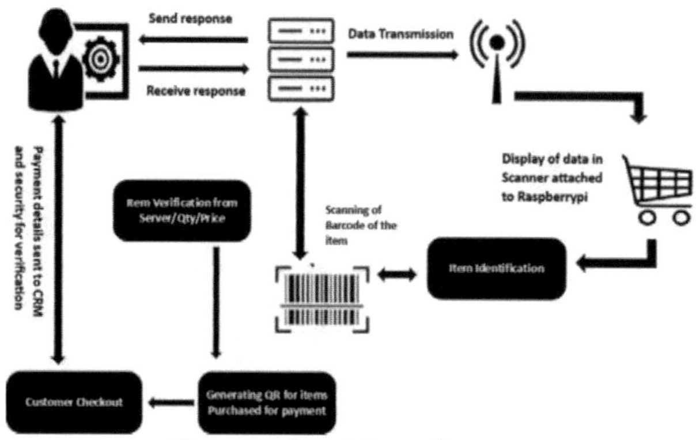

Figure 3. Workflow diagram.

The circuit diagram consists of Raspberry Pi-4 as the main component to which the buzzer, GM-73 Barcode/QR scanner is connected. The pins used for connecting the GM73 module are 5V, GND, RX, and TX (figure 4).

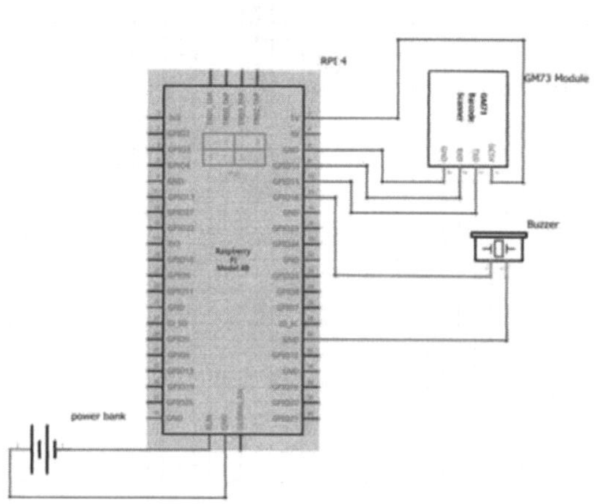

Figure 4. Circuit diagram.

The Pins used for the buzzer are the GPIO input pin and ground. The Raspberry Pi is powered through a type C port from a 5V, 3A Power bank which is depicted as RUN and GND in the circuit diagram. Note that the LCD is also connected to the Raspberry Pi externally through HDMI and USB ports which are not specified as Pins in the Pinout Circuit diagram

5. Result and discussion

The analytical report derived from the perspective of customers on waiting times in supermarkets demonstrates a clear and significant improvement in billing wait times after the implementation of the proposed system. A comparison with the previous/existing method reveals that the wait time in the proposed model, as depicted in figure 5, is only half of the wait time in the previous approach. The data indicates a substantial reduction in waiting times with the adoption of the proposed system.

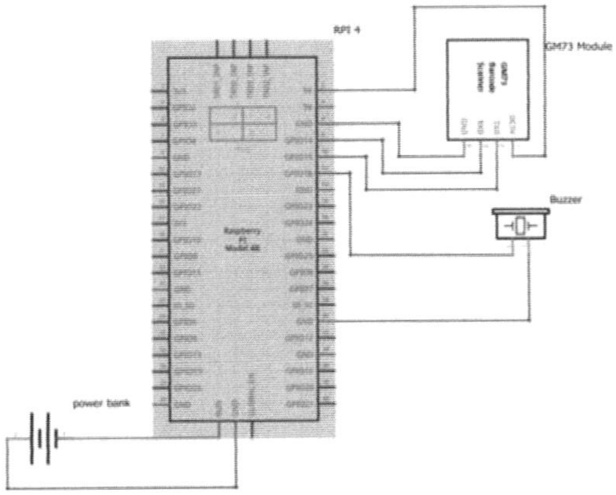

Figure 5. Circuit diagram.

The average waiting time per 10 Customer's Ratio there is the differentiation between existing and proposed models is about 4 mins delay can be avoided in approx. The results of implementing the automated shopping trolley system were highly promising, demonstrating its effectiveness in transforming the shopping experience. The barcode scanning feature of the system consistently read barcodes accurately and swiftly retrieved product information. This streamlined the product selection process, enabling customers to scan the items quickly and efficiently they intended to purchase. Users appreciated the system's responsiveness and accuracy in capturing the correct product details. The integration of the touchscreen LCD display and user- friendly GUI received positive feedback from users. The intuitive interface allowed customers to easily navigate through the system, self-scan products, and make modifications to their purchase lists. The real-time display of the running bill total provided transparency and enabled customers to stay informed about their expenses. The wireless smart device, connected to the supermarket's backend database, ensured up-to-date and reliable product information. Customers benefited from real-time updates on product prices, stock availability, and expiry dates, allowing them to make informed decisions while shopping. The system's ability to notify customers about expired products proved to be an invaluable feature, ensuring their safety and preventing the purchase of expired items (figure 6).

The implementation of the security page dedicated to theft detection received positive reviews. Security personnel found the page helpful in cross-checking the purchased items during the checkout process, minimizing the risk of theft and unauthorized product removal. This additional layer of security instilled confidence in both the supermarket staff and customers, ensuring a secure shopping environment.

Throughout the testing phase, the system demonstrated stability and reliability. It operated smoothly with minimal disruptions or technical issues. The meticulous testing and optimization procedures performed prior to implementation played a crucial role in ensuring the system's robustness and overall performance.

User feedback and observations during testing validated the success of the implemented system in improving shopping efficiency, convenience, and security. Customers reported a more streamlined and convenient shopping experience, with reduced waiting times at the checkout counter. The system effectively met the objectives and requirements set forth in the paper.

Continued efforts will be directed toward ongoing evaluations and refinements to address any identified limitations and further enhance the system's performance. Regular updates and improvements will ensure that the automated shopping trolley system remains efficient, reliable, and capable of delivering an enhanced shopping experience for the customer.

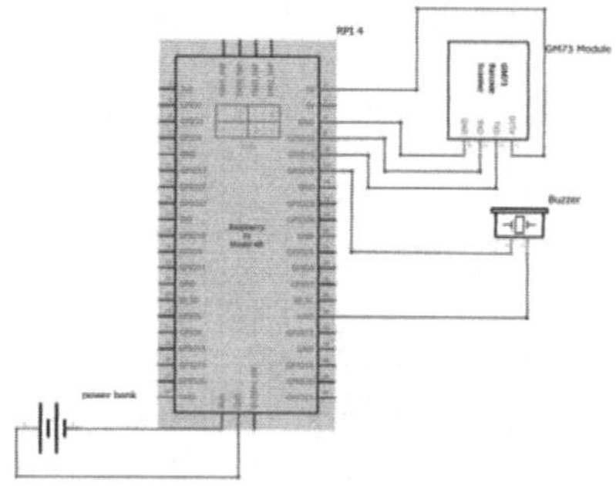

Figure 6. Working model.

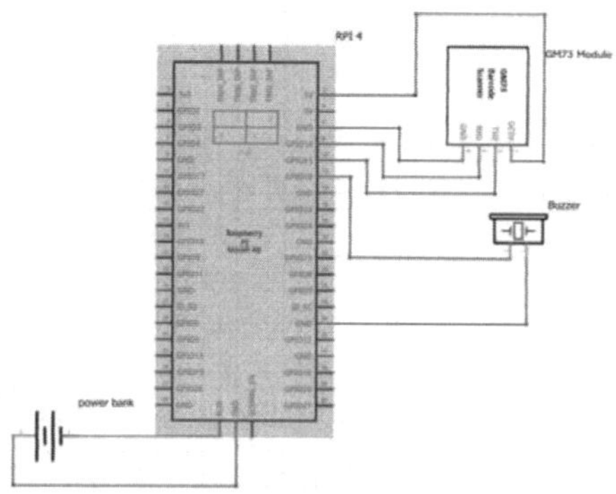

Figure 7. GUI for user interaction.

The customer interaction includes a Graphical User Interface (GUI) where customers can conveniently scan products by tapping the "Scan Product" button (figure 7). Each item is added individually, and if a customer wants to remove an item, they can simply tap on it and select the "Delete" option. To proceed with the payment, customers can tap on the "Pay" button, which will generate a QR code for seamless transaction processing. A notable feature of the system is its ability to detect expired products. If a customer unintentionally scans an expired product, the system will emit a buzzer sound as a warning, and the QR code won't be generated for payment.

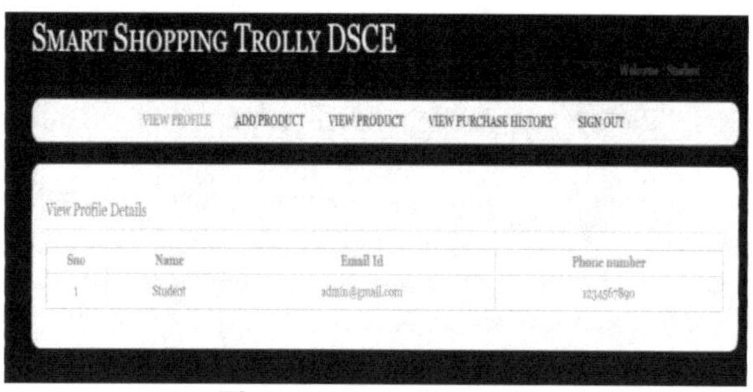

Figure 8. Admin-page.

It consists of two pages: the admin page (figure 8) and the Security page (figure 9). The admin page provides access to the administrator, allowing them to add products by clicking the "Add Product" button. The added products are then stored in the database. By clicking on the "View Product" button, the administrator can see all the products in the database, along with their quantity and price (figure 11). Additionally, the administrator can access the purchase history of customers by clicking on the "View Purchased History" button (figure 12). The Security page also has access to the purchase history, which is helpful for security personnel during checkout. The purchase history displays the date, time, and name of the customer for each transaction.

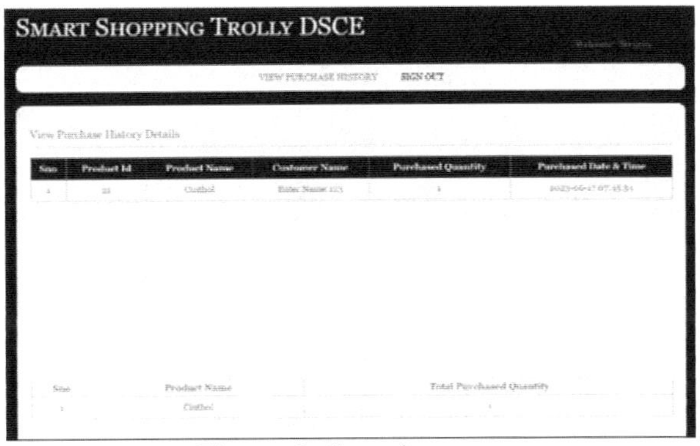

Figure 9. Security page.

The journey towards optimal system evolution necessitates a keen focus on technical compatibility. As the system evolves, integration with new technologies and platforms might be required. This demands a proactive approach in designing the system's architecture to remain adaptable and open to future enhancements. Emphasizing the need to establish flexible interfaces and communication protocols underscores the commitment to accommodating emerging technologies seamlessly. Achieving a robust and reliable system evolution involves addressing potential data disruptions. Given the reliance on wireless communication and data exchange, factors like network stability and security vulnerabilities must be thoroughly addressed. Continuous monitoring, implementing fail-safe mechanisms, and adopting robust encryption strategies are vital to mitigate data disruption risks and sustain the system's integrity (figure 10).

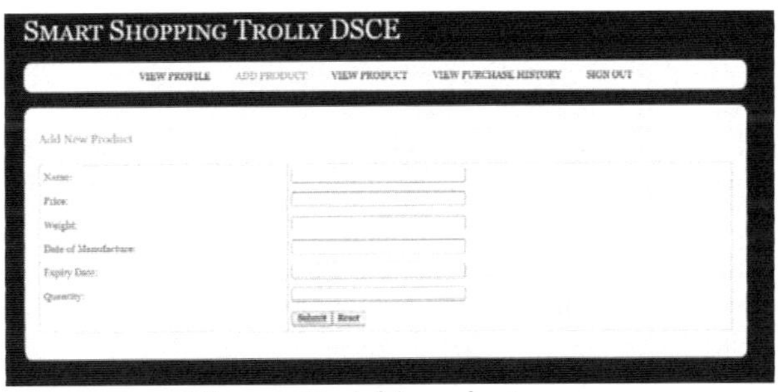

Figure 10. Add product page.

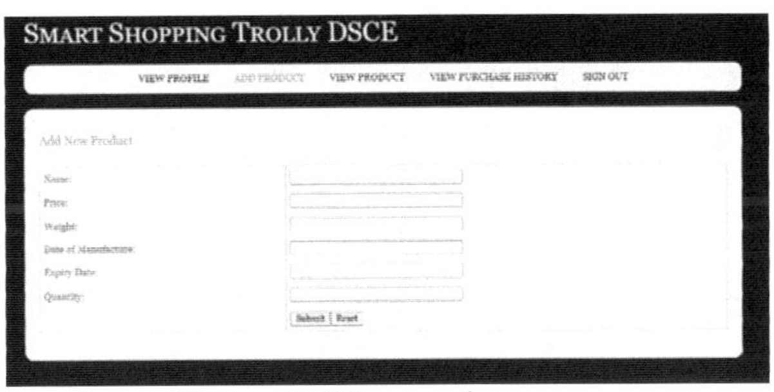

Figure 11. View products page.

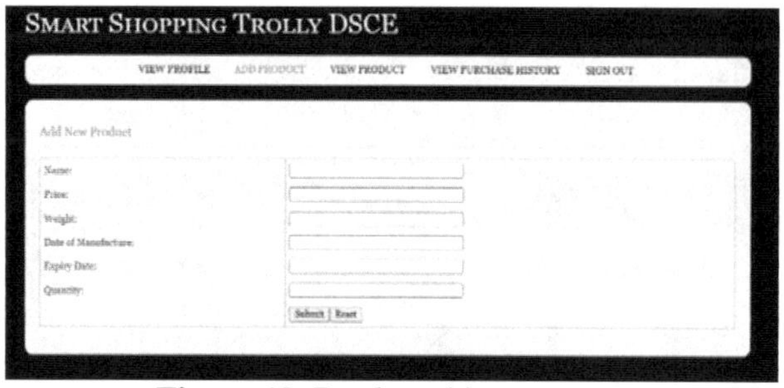

Figure 12. Purchase history page.

6. Conclusion

This In conclusion, the implementation of the automated shopping trolley system has proven to be a valuable solution for improving the shopping experience. With features such as barcode scanning, touch screen display, and integration with a backend database, the system enhances efficiency and convenience for customers. The inclusion of a security page further enhances security measures. Positive user feedback validates the system's effectiveness in delivering a streamlined and secure shopping experience. Ongoing efforts will continue to refine and optimize the system, ensuring its continued success in meeting the evolving needs of customers. Overall, the automated shopping trolley system offers significant benefits for both customers and retailers alike.

The envisioned future for the innovative smart shopping trolley with automated billing holds immense promise across various dimensions. Seamless integration with dedicated mobile applications is poised to revolutionize the shopping experience, offering users personalized shopping lists, preferences, and payment methods directly synced with the smart trolley. The trolley's advanced AI capabilities will not only analyze shopping patterns and trends, enabling accurate product recommendations and predictive shopping lists, but also offer real-time insights into available discounts and promotions. Augmented reality will be harnessed to provide shoppers with an interactive and informative in- store navigation system, guiding them to desired products, highlighting relevant information through the trolley's interface, and suggesting complementary items.

Moreover, this innovative trolley will feature automated restocking functionality through integrated sensors, which will detect dwindling supplies and intelligently add items to the shopping list, ensuring popular products are consistently available. Biometric authentication methods such as fingerprint and facial recognition will guarantee secure access to the trolley and authorized payments, upholding convenience and security.

The envisioned future for the innovative smart shopping trolley with automated billing holds immense promise across various dimensions. Seamless integration with dedicated mobile applications is poised to revolutionize the shopping experience, offering users personalized shopping lists, preferences, and payment methods directly synced with the smart trolley. The trolley's advanced AI capabilities will not only analyze shopping patterns and trends, enabling accurate product recommendations and predictive shopping lists, but also offer real-time insights into available discounts and promotions. Augmented reality will be harnessed to provide shoppers with an interactive and informative in- store navigation system, guiding them to desired products, highlighting relevant information through the trolley's interface, and suggesting complementary items.

7. References

[1] Jaishree M, Lakshmi Prabha K R, Jeyaprabha S and Mohan K 2021 *2021 7th International Conference on Advanced Computing and Communication Systems, ICACCS* p 20800057

[2] Zhang H, Shi G, Liu L, Zhao M and Liang Z 2019 *2018 8th International Conference on Image Processing Theory, Tools and Applications, IPTA 2018 – Proceedings* p 18403560

[3] Visalatchi A R, Navasri T, Ranjanipriya P and Yogamathi R 2020 *Proceedings of the 4th International Conference on Computing Methodologies and Communication, ICCMC* p 19569433

[4] Ezhilazhagan C, Adithya R, Burhanuddin Y L and Charles F 2017 *IEEE International Conference on Recent Trends in Electronics, Information and Communication Technology, RTEICT* p 16582902
[5] Hashim N M Z, Ibrahim N A, Saad N M, Sakaguchi F and Zakaria Z. 2013 *International Journal of Emerging Trends & Technology in Computer Science (IJETTCS)* **2** 78
[6] Badi R and Momin B 2018 *2018 3rd International Conference for Convergence in Technology, I2CT* p 18233213
[7] Gupta S, Kaur A, Garg A, Verma A, Bansal A and Singh A 2013 *2018 3rd International Conference for Convergence in Technology, I2CT* p 3083
[8] Yewatkar A, Inamdar F, Singh R, Ayushya and Bandal A 2016 *Procedia Computer Science* **79** 793
[9] Karjol S, Holla A K and Abhilash C B 2018 *Communications in Computer and Information Science* **801** 373

A Review on Deep Learning based Algorithms for Video Super Resolution

Deekshitha Arasa[1*], Sivaramakrishnan S[2], and Sneha Sharma[3]

[1] Department of Electrical and Electronics Engineering, Dayananda Sagar College of Engineering, Bengaluru, Karnataka, India
[2] Department of Information Science and Engineering, New Horizon College of Engineering, Bengaluru, Karnataka, India
[3] Department of Electronics and Communication Engineering, Dayananda Sagar University, Bengaluru, Karnataka, India

deekshitha-eee@dayanandasagar.edu

Abstract. A HR (High Resolution) video frame may be created from a set of LR (Low Resolution) video frames using a process called Video Super Resolution (VSR). Deep Learning (DL) techniques have surpassed conventional techniques in terms of efficiency. To achieve video super-resolution, the inter-frame information must be extracted. The numerous categories for possible deep learning techniques include methods with alignment and methods without alignment, for example. Details about the different approaches' architecture designs and implementation are provided in this review paper. Lastly, a comparison and summary are given for the various methods based on some standard datasets is provided.

1. Introduction

Today, one of the most prevalent forms of multimedia is video. A single image is dealt with by the image super-resolution process at a time [1], whereas VSR methods deal with numerous frames or images and attempt to determine the relationship between each frame. Each frame will be at a different location in a video, or they are related in time. From frame to frame, various items' positions shift. Object movement slows down if distinct frames are not properly aligned. Algorithms for image super-resolution can process each frame if VSR is viewed as an extension of that technology. But the effectiveness of such algorithms might not be adequate.

In light of this, numerous Video super-resolution algorithms have been put forth recently. Traditional approaches and deep learning methods are two major groups into which they can be divided. These methods of upscaling make an effort to make use of some natural preferences and accurately predict the mobility between the frames. The LR frame is first converted to the frequency domain in a traditional procedure. After being estimated in this domain for the high-resolution frame, the outcome frame is translated once more into the spatial domain. Some techniques employ the FT (Fourier Transform), which aids in extending the spectra of the collected signal and so improving resolution. However, these strategies show to be less effective than deep learning methods [2].

Deep learning-based super-resolution approaches were presented recently as a result of deep learning's enormous success across all sectors. For feature extraction, inter-frame alignment, and feature fusion, techniques like RNN, CNN, and GAN (Generative Adversarial Network) use a significant number of

HR and LR video sequences as input. These techniques then generate the HR video sequences for the associated LR sequences.

The paper is organized as follows: Section II discusses the video super resolution techniques. It is classified as 1. Methods with alignment and 2. Methods without alignment.

Various algorithms under each category are discussed. Mathematical modelling of these algorithms along with their architectures are mentioned. Finally, the performance parameters of each algorithm are compared in section 3. Section 4 has the conclusion.

2. Video super resolution methods

The approaches for VSR draw inspiration from the single-image super-resolution techniques already in use because videos are recordings of moving visual sound and images. A single image super resolution's objective is to convert a low into a high-resolution image. As opposed to Video Super quality, which aims to convert low-quality videos to high-definition ones. Here, there is a temporal connection between the various video frames. It indicates that the positions of several items vary from frame to frame. Aligning objects is therefore necessary. There are two widely used methods for aligning various video frames. Deformable Convolution (DC) and MEMC ("Motion Estimation and Motion Compensation") are the names of these techniques [3,4].

As video frames are explicitly aligned, we divide the current techniques into 2 main categories: approaches with alignment and approaches without alignment as shown in figure 1. The approaches will be described in depth in the sections that follow.

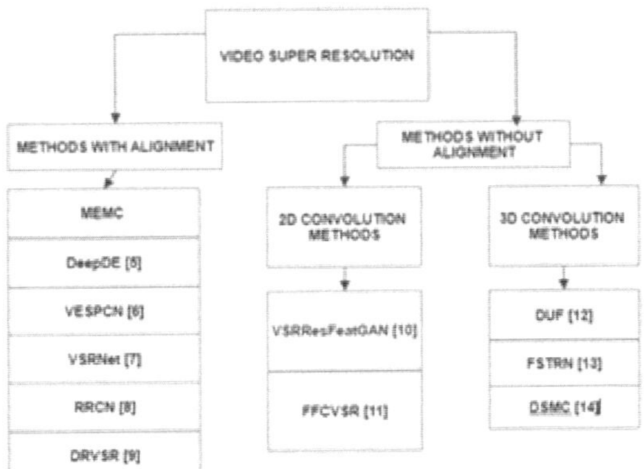

Figure 1. Categorization of various existing super-resolution methods.

Other modules used for feature extraction, reconstruction, and fusion are not taken into consideration because all approaches are categorized as per frames and are "explicitly aligned" and which approaches are primarily employed for alignment.

2.1. Methods with alignment

By employing the retrieved motion data before the following reconstruction, the approaches with alignment explicitly align nearby frames with the target frame. These methods mostly employ deformable convolution or MEMC, which are 2 widely used methods for aligning frames.

Motion estimation is utilized to extract interframe information relate to motion, whereas motion compensation is utilized to distort frames in accordance with interframe motion info and to align one frame to another.

The optical flow approach is used to accomplish most motion estimating algorithms. This method makes an effort to determine the mobility between two adjacent frames using their temporal domain correlations and variances. An optical flow technique typically requires 2 input frames, such as Ii and Ij. The target frame is the first, while the adjoining frame is the second. The approach then uses the following formula to compute a vector field of "optical flow" Fij from frame Ii into Ij as represented in equation (1).

$$F_{i\to j}(h_{i\to j}, v_{i\to j}) = ME(I_i, I_j; ME) \quad (1)$$

Here hi→j and vi→j indicate horizontal & "vertical components of Fi→j, ME(·) represents the function applied to determine optical flow, and ME represents its parameter. Motion compensation is applied to conduct image transformations between images based on motion info to align neighbouring frames along with the target frame". Generally, a "compensated frame" Ij is represented as shown in equation (2):

$$I_j = MC(I_i, F_{i\to j}, MC) \quad (2)$$

Here MC(·) indicates a function of motion compensation, Ii, Fi→j, as well as MC represent neighbouring frame, optical flow along with parameter. MC may be obtained using a variety of techniques, including STN ("Spatial Transformer Networks") and bilinear interpolation.
The processes of MC and ME could both be carried out using DL or a conventional (non-deep learning) manner. We further subcategorize the MEMC approaches into 2 groups based on whether conventional or deep learning techniques were utilized to apply to MC or ME. If any of the procedures in ME or MC make use of a DNN, the procedure comes under the DL umbrella; otherwise, the procedure goes under the conventional umbrella.
The methods with alignment algorithms will be described in depth in the sections that follow.

2.1.1. Deep-DE. The mathematical modeling of Deep DE involves the integration of differential equations and neural networks. The first step in mathematical modeling is to define the differential equations that represent the problem you want to solve or model. These can be ordinary differential equations (ODEs) or partial differential equations (PDEs), depending on the system's dynamics and spatial dimensions. An ordinary differential equation can be represented as:

$$\frac{dy}{dt} = F(y,t) \quad (3)$$

Where y is the state vector of the system, t is the time variable and F(y,t) is the function defining system dynamics. Similarly, a partial differential equation can be represented as:

$$\frac{\partial u}{\partial t} = \nabla_u^2 + F(u, \nabla_u, t) \quad (4)$$

Where u is the dependent variable, is the Laplacian operator and represents the systems partial derivatives w.r.t u, and t respectively. In Deep DE, neural networks are used to approximate the solution to the differential equations. The architecture of the neural network can vary depending on the problem and complexity of the equations. The two fundamental parts of the Deep Draft-Ensemble (Deep-DE) learning approach are the production of the SR draft ensemble and its optimal reconstruction. The typical feedforward reconstruction pipeline needs to be updated to significantly improve its capacity to compute various super-resolution outcomes while taking high motion variation and potential process faults into account. Then, in a deep convolutional neural network (CNN), SR drafts are mixed via a nonlinear method. A CNN is used for feature extraction, super-resolution, and fusion. It receives both the bicubic-interpolated target frame of LR and the SR draughts. The CNN in Deep-DE contains 4 convolutional layers, as illustrated in figure 2 below: the first 3 are conventional convolutional layers, while the fourth layer is a "deconvolution layer". By using a non-iterative process, the goal of this strategy is to reduce the majority of the calculation. This approach regularly outperforms alternative approaches on both synthetic and real-world data, avoiding parameter adjustment during the test phase [5]. It is noted that SR drafts created by straightforward feedforward reconstruction techniques by adjusting motion estimate settings, often contain enough information for estimating the final HR image. Based on this discovery, CNN was used to combine the procedures of deconvolution and reconstruction.

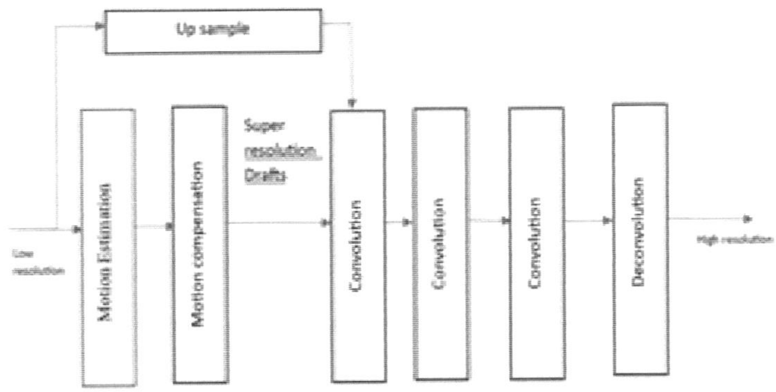

Figure 2. Architecture of Deep-DE.

The sizes of the kernel are 1×1, 3×3, 11×11, and 25×25, and numbers of channels are 512, 1, 256, and 1.

2.1.2. VSRNET. The network architecture of VSRNET, which is depending on the SRCNN image super-resolution technique, is shown in figure 3.

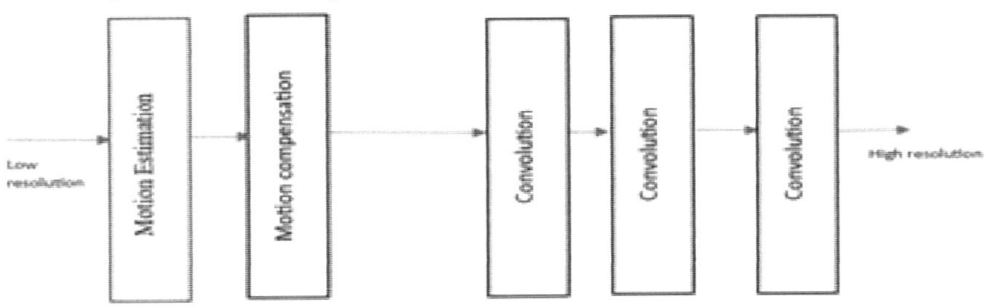

Figure 3. VSRNET architecture.

The basic components of VSRnet are three convolutional layers, three MEMC modules, and 3 rectified linear units (ReLUs), except for the final convolutional layer [6]. The quantity of input frames distinguishes SRCNN from VSRnet the most. In contrast, VSRnet employs a series of subsequent, corrected frames, whereas SRCNN only accepts a single input frame. The Druleas algorithm computes the movement information between frames. Additionally, the FSE ("Filter Symmetry Enforcement") system and the "adaptive motion compensation" method that VSRnet suggests are utilized separately to speed up training and lessen the effect of erroneous compensated frames and, as a result, could enhance video super-resolution performance. With the aid of pretraining, filter symmetry enforcement, motion-compensated input frames, and it was able to enhance the quality of reconstruction and decrease the training time.

2.1.3. RRCN. Residual and Recurrent Convolutional Networks (RRCN) are a type of neural network architecture that combines the concepts of residual networks (ResNet) and recurrent networks. The mathematical formulation of a residual block can be represented as follows: Given an input feature map X and a function F(X) representing the convolutional layers, the output of a residual block with skip connection can be written as:

$$Output = X + F(X) \tag{5}$$

Where Output is the final output of the block. Recurrent Neural Networks (RNNs) are designed to handle sequential data by incorporating feedback loops that allow information to persist. The Long Short-Term

Memory (LSTM) and Gated Recurrent Unit (GRU) are popular types of RNNs. The mathematical formulation of an LSTM cell can be represented as:

$$h_t, c_t = LSTM\left(x_t, h_{t-1}, C_{t-1}\right) \tag{6}$$

Where is the input of time step t, hidden state at time step t-1 and ct is the cell state at time step t. The RRCN combines the concepts of ResNet and recurrent networks. One way to design an RRCN is to use residual connections within the recurrent layers. In this case, the mathematical formulation of an RRCN can be represented as:

$$h_t, c_t = LSTM\left(x_t, F(x), h_{t-1}, C_{t-1}\right) \tag{7}$$

Where X is the input feature map, F(X) represents the convolutional layer, LSTM(.) is the LSTM cell. As depicted in figure 4, the RRCN is a bidirectional RNN that learns a residual image [7]. RRCN suggests an unsynchronized full RNN with numerous "consecutive video frames" as input and only the middle one being super-resolved.

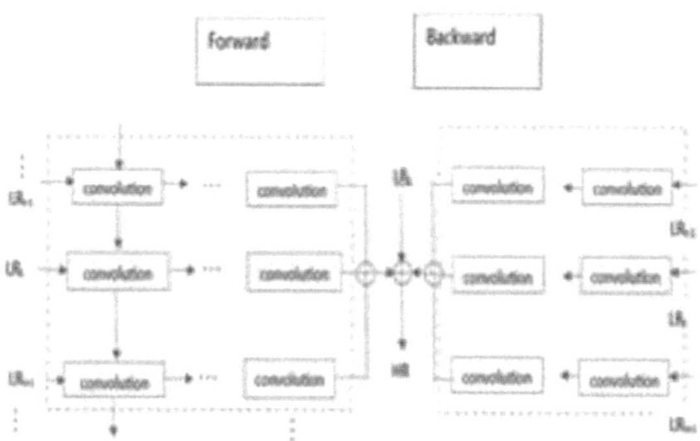

Figure 4. Network architecture of RRCN.

For MEMC for the target frame and its neighboring frames, RRCN employs the "combined local-global with total variable (GLG-TV)" approach. The network receives the adjusted frames as input. The O/P's of the forward as well as backward network, respectively, are combined after the forward & recurrent convolution have been applied. The target frame is then added to the input, and the outcome is then obtained. The self-ensemble strategy is also used by RRCN to further enhance performance, and it is combined with the results of the single image super-resolution approach, EDSR+, to create the 2 models known as RRCN+ & RRCN++, respectively. Table 1 shows that the PSNR values drop dramatically in the absence of MES and Res. Therefore, the greatest performance requires both MES and Res.

Table 1. The PSNR average values for the testing dataset of VideoSet4 for ×3 with various parameters.

Setting	PSNR (Peak Signal to Noise Ratio)	SSIM (Structural Similarity Index)
Without Motion Compensation (MC) and Residual Connections (Res)	26.81	0.8183
Without MC, Res	27.12	0.8322
With MC, without Res	27.36	0.8455

With MC, RES	27.65	0.8557"

2.1.4. VESPCN. A spatial MCT ("Motion Compensation Transformer") module is proposed by the VESPCN ("Video Efficient Sub-Pixel Convolutional Network") for motion estimation and compensation [8]. To extract and combine features, the compensation frames are then input through a succession of convolutional layers as depicted in figure 5.

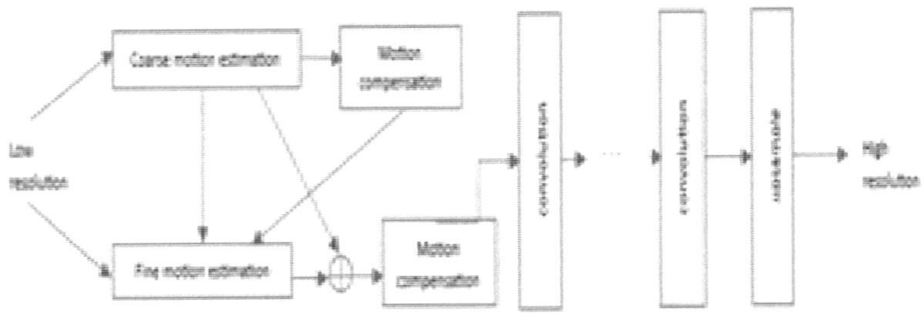

Figure 5. Architecture of VESPCN.

MCTs are used to extract motion data and compensate for motion. First, the network uses the target frame and a surrounding frame as inputs during the coarse estimation stage. The sub-pixel convolutional layer is one of five convolutional layers that make up the coarse network. Additionally, to obtain rough findings for optical flow estimate, it first executes the 2 down sampling operations twice before doing the 4 up-sampling operations via a sub-pixel "convolutional layer". the optical neighbouring frame is warped second in accordance with the optical flow. The target frame, adjacent frame, flow calculated during the coarse estimate stage, and the warped neighbouring frame are the inputs of the fine network, which has an architecture similar to that of the coarse network. To achieve the fine optical flow, it first does a ×2 down sampling followed by a ×2 up sampling at the network's end. The fine optical flow is applied in conjunction with the coarse optical flow to determine the final estimation outcome. Finally, the last optical flow warps the adjacent frame once more to align the warped frame with the target frame. A new sub-pixel convolution layer may super-resolve LR data to HR space with a very small increase in computational cost than in the deconvolution layer during training time.

2.1.5. DRVSR. As per predicated optical flow information, the DRVSR ("Detail-Revealing Deep Video Super-Resolution") technique suggests a SPMC ("Sub-Pixel Motion Compensation") layer that may carry out the up sampling as well as motion compensation operations concurrently for nearby input frames [9]. Figure 6 depicts the DRVSR network architecture.

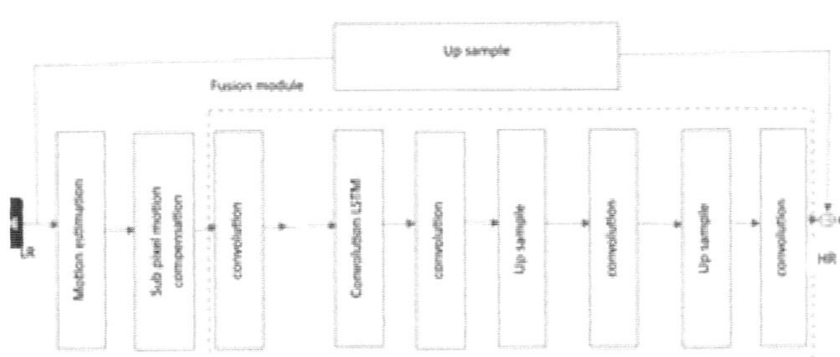

Figure 6. Architecture of DRVSR.

Motion estimation, motion compensation utilizing the layer of SPMC, and fusion are the three major modules of DRVSR. Using the MCT (Motion Compensation Transformer) network, the motion prediction module is implemented. The sampler and grid generator are the two sub-modules that make up the SPMC layer. According to the optical flow, the grid generator 1st converts the coordinates within LR space to the coordinates within HR space. The sampler then completes the operation of interpolation within the HR space. The fusion module employs stride 2 convolutions to accomplish down-sampling before performing deconvolution to perform up sampling to retrieve the HR residual picture of the target frame. This residual image produces the end output when combined with the up-sampled LR target frame. To manage spatiotemporal data, DRVSR additionally uses the ConvLSTM module.

2.2. Methods without alignment

The nearby frames are not aligned using these techniques for video super-resolution [10]. These methods use spatio or spatial-temporal information primarily for feature extraction. These techniques are further divided into four categories: 2D conv ("2D convolution") methods, RCNN, three-dimensional convolution, and non-local network-based approaches.

To conduct operations like feature extraction, fusion, and super-resolution spatially, the input frames are immediately fed into a two-dimensional convolutional network rather than alignment processes like MEMC between frames. These techniques develop the correlation knowledge within the frames on their own. The two representative methods are VSRResFeatGAN and FFCVSR.

2.2.1. VSRResFeatGAN. The Generative Adversarial Network (GAN) is used in this method to handle VSR tasks [10]. Convolutional layers and residual blocks make up the generator in figure 7. Then, each residual block is followed using the ReLU activation function and is made up of two convolutional layers. Three convolutional groups—one each for a convolutional layer, BN (Batch Normalisation), as well as LeakyReLU—plus a fully connected layer, make up a discriminator.

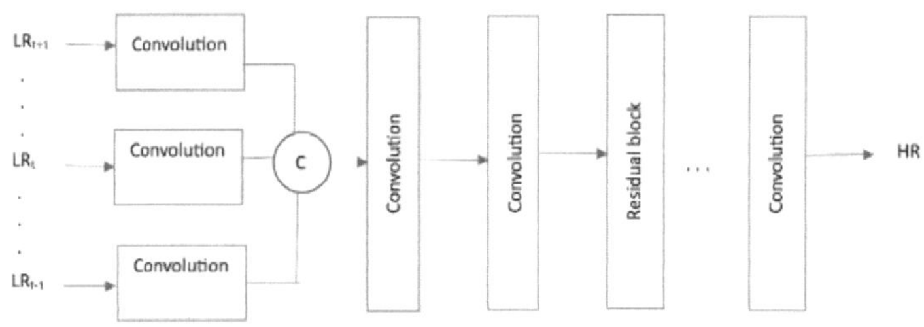

Figure 7. VSRResFeatGAN network architecture.

Table 2. The Myanmar test sequences' average PSNR and SSIM values for a scale factor of three.

	VSRNet	VSRResNet	VSRResFeatGAN
PSNR	34.42	35.86	28.53
SSIM	0.9247	0.9478	0.9216

According to table 2's findings, when a VSRResNet model (deeper model) is trained, the PSNR significantly rises by 1.6 dB in comparison to the VSRNet model. However, using adversarial and feature-space losses, which are employed in VSRResFeatGAN, results in a drop in the PSNR from 35.86 dB to 28.53 dB while still enhancing picture quality. These losses deviate from the conventional pixel-wise difference.

2.2.2. FFCVSR. Figure 8 depicts the design of the frame as well as FFCVSR (Feature Context Video Super Resolution). It uses inter-frame data in a different way [11] and is made up of many local networks and context networks. The network accepts the LR misaligned video frames and the previous frame's HR output as direct I/Ps to recover high-frequency information and preserve temporal consistency.

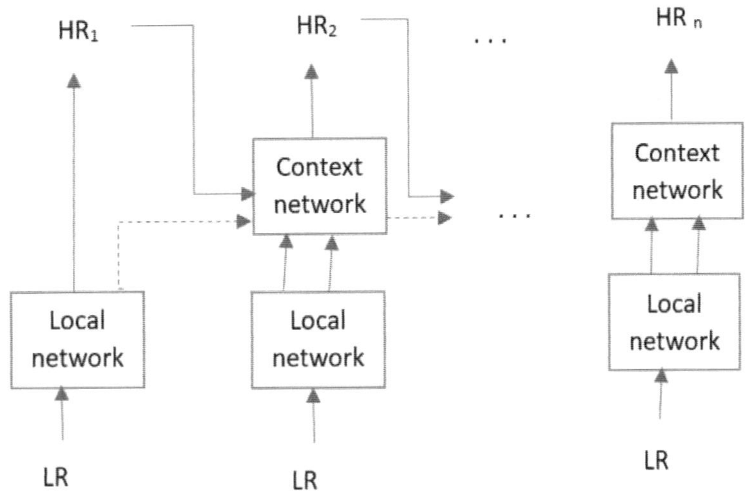

Figure 8. Architecture of FFCVSR.

A fully end-to-end trainable framework made up of a local network and context network is suggested to simultaneously use previously inferred frames and features rather than only leveraging numerous LR frames to independently construct each output frame. Additionally, a suppression-updating method is built based on the framework's properties to successfully address the issue of error accumulation in high-frequency information. Numerous tests, including an ablation study, show that this method greatly improves upon the state-of-the-art on a common benchmark dataset and is effective in delivering high-quality temporal consistency video resolution enhancement. Let's consider a video sequence with low-resolution frames as input, denoted as LR_frames. The goal is to generate a corresponding video sequence of high-resolution frames, denoted as HR_frames. Use a pre-trained deep neural network, such as a convolutional neural network (CNN), to extract high-level features from LR_frames. This network is often referred to as a feature extractor or backbone network.

Apply a feature extraction function to each frame in LR_frames, denoted as extract_features(frame). This function extracts relevant spatial features that capture the content information in the low-resolution frame. Define a context window size W that determines how many neighboring frames to consider for each frame in the video sequence. For each frame frame_t in LR_frames, select W neighbouring frames before and after frame_t to form a temporal context window context_window_t. The size of context_window_t will be 2*W + 1. For each frame frame_t in LR_frames, extract features using the feature extraction function: features_t = extract_features(frame_t).

For each frame frame_t in LR_frames, extract features from its temporal context window: context_features_t = [extract_features(frame) for frame in context_window_t]. Use an upscaling network, such as another CNN or sub-pixel convolutional layers, to upscale the features features_t to high-resolution features HR_features_t.

The upscaling network takes into account the information from both features_t and context_features_t to produce more accurate high-resolution representations. Use an inverse function to map HR_features_t back to high-resolution frame HR_frame_t, denoted as reconstruct (HR_features_t).

Reconstruct high-resolution frames for all frames in LR_frames using the upscaling and reconstruction steps. Define an objective function that quantifies the difference between the generated high-resolution frames HR_frames and the ground truth high-resolution frames GT_frames.

The objective function can be a combination of various loss terms, such as pixel-wise mean squared error, perceptual loss, and temporal consistency loss. Minimize the objective function using gradient-based optimization methods (e.g., stochastic gradient descent, Adam) to update the parameters of the feature extractor, upscaling network, and other components.

By incorporating feature context from neighbouring frames, Feature Context Video Super Resolution can effectively enhance the resolution of video frames, capturing both spatial and temporal information

IOP Conf. Series: Materials Science and Engineering 1295 (2023) 012009 doi:10.1088/1757-899X/1295/1/012009

for improved reconstruction quality. Keep in mind that the specific architecture and details of FCVSR may vary based on the research papers or implementations in the literature. While 3D convolution works in the spatiotemporal domain, 2D convolution solely uses spatial information via sliding of the kernel over the input frame. Due to the consideration of frame correlations through the extraction of temporal information, this is advantageous for the processing of video sequences. DUF, FSTRN, and 3DSRnet are some examples of approaches in this area.

2.2.3. DUF. The "dynamic up sampling filters" (DUF) are based on dynamic filter networks, which may produce appropriate filters for certain inputs and then use those filters to produce corresponding feature maps [12] (figure 9). The filtering and up-sampling operations are both carried out by DUF.

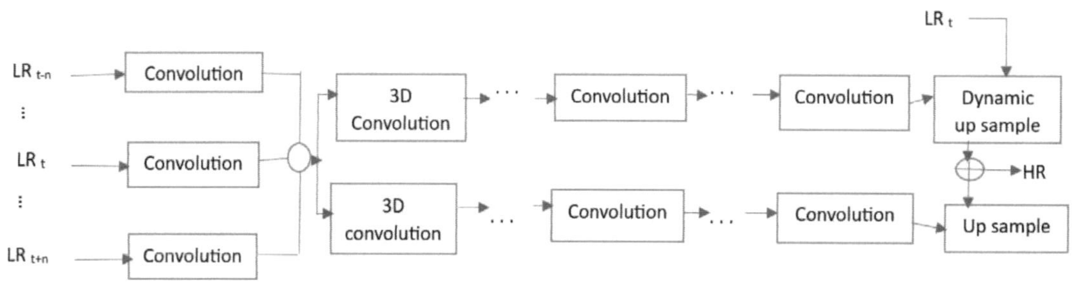

Figure 9. Architecture of DUF.

This approach was tried using a basic network of 16 layers, followed by 28 layers, and finally 52 layers. Increasing the number of layers in this algorithm produced better outcomes for difficult scenarios. Using a NVidia GeForce GTX 1080 Ti, it took roughly two days to train the network, and the runtimes for producing a single 1920×1080 O/P frame from seven 480 ×270 input frames are 2.8189s 0.8382s, and 0.4030s, for 52L, 28L, and16L, respectively. In the inference phase, local filtering takes up around half of the runtime.

2.2.4. FSTRN. To extract data from successive frames, the FSTRN ("Fast Spatio-Temporal Residual Network") performs a factorized 3D convolution. [13], as seen in figure 10.

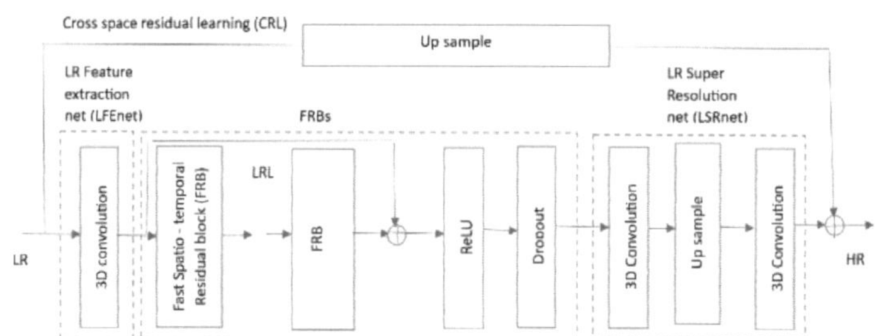

Figure 10. Architecture of FSTRN.

In FSTRN, a k×k×k three-dimensional convolutional kernel is divided into two cascaded kernels, each of which has a size of 1×k×k as well as k×1×1. This reduces computation by employing the three-dimensional convolution directly.

FSTRN is composed of FRBs ("Fast Spatiotemporal Residual Blocks"), LR feature fusion and up-sampling SR net (LSRNet), LFENet ("LR Video Shallow Feature Extraction Net"), as well as GRL ("Global Residual Learning") module. LFENet employs three-dimensional convolution to extract features for input frames of LR.

2.2.5. DSMC. A “Dual Subnet and Multi-Stage Communicated Up sampling” (DSMC) is a deep neural network that is used for super video resolution with large motion [14]. The design is revealed in figure 11.

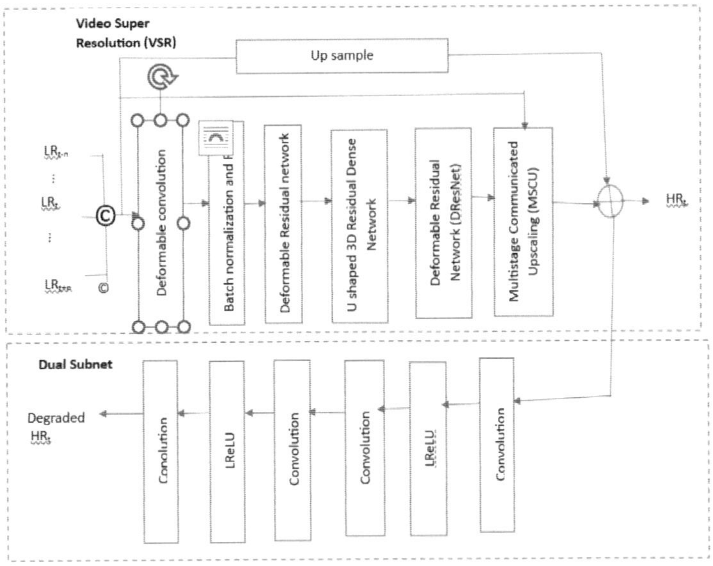

Figure 11. Architecture of DSMC.

It creates a U3D-RDN (“U-shaped residual dense network with 3D convolution”) for both coarse and fine implicit MEMC feature extraction. To fully utilize the intermediate up sampling findings for directing the VSR, DSMC has developed a new MSCU (“Multi-Stage Communicated Up Sampling”) module. In addition, “a dual subnet is designed to facilitate the training of the DSMC, whose dual loss aids in reducing the solution space and improving generalization ability.

For coarse feature extraction, DSMC first applies deformable convolution on the input consecutive frames. To extract fine spatial information, a deformable residual network (DResNet) is used to process the output feature maps. Following that, the feature maps are” sent into U3D-RDN for dimension reduction as well as correlation analysis of spatiotemporal features. The feature maps are delivered to the module of MSCU after another DResNet module. The super-resolved HR frames are then produced by DSMC using a dual subnet for training. It should be emphasized that the DMC's loss computation only uses the dual and VSR subnet's results [15].

Briefly stated these three-dimensional convolutional algorithms may extract spatio-temporal correlations from a sequence of frames as opposed to doing motion prediction to extract motion information from a sequence of frames as well as motion compensation to align them. Nevertheless, the majority of the algorithms are constrained for real-time video tasks of super-resolution due to their increased computational complexity when compared to 2D convolutional methods [15].

3. Performance comparisons

Both alignment and non-alignment methods have their strengths and weaknesses, and the choice between them depends on the specific requirements of the application. Alignment-based methods tend to produce more visually consistent results, especially in videos with complex motion, while non-alignment methods are computationally more efficient but may not fully utilize the temporal information in the video. Researchers often explore a combination of these approaches to strike a balance between accuracy and efficiency, leading to innovative solutions in the field of video super-resolution.

As evaluation measures, the video super-resolution models are compared using SSIM and PSNR. The highest value of the pixel and the MSE (“Mean Squared Error”) between the frames are used to define PSNR. The images or frames’ structural similarity is to be measured using SSIM. Gaussian blurring with downsampling (BD) and bicubic downsampling with image- resize function (BI) are the two types of degradation [15].

Table 3. Comparison of the approaches on the datasets with scale factor × 4.

	Bicubic downsampling with Image-resize function (BI)		Gaussian blurring and downsampling (BD)		Advantages
	PSNR	SSIM	PSNR	SSIM	
Deep- DE	-	-	29.00	0.8870	1) SR draft ensemble generation: Upgrade the existing pipeline for feed-forward reconstruction; Improve the capability to compute various super-resolution findings; significant motion variation and various latent artefacts are taken into account. 2) CNN is used to combine the deconvolution and reconstruction phases in SR draft-based optimal reconstruction, avoiding parameter adjustment in the test phase.
VSRNet	24.84	0.7049	-	-	1) To improve the spatial resolution, it was suggested to use a CNN that was trained on spatial & temporal regions. 2) To increase accuracy and efficiency, pre-training techniques, filter symmetry enforcement, and motion-compensated frames are all combined. 3) To manage motion blur and quickly moving objects, an adaptive motion correction technique was used.
VESPCN	31.85	0.8834	-	-	1) Developing a motion compensation strategy on the basis of a quick multi-resolution spatial transformer module. 2)The precision and temporal consistency of VSR is improved in real-time using a spatio-temporal sub-pixel convolution network.
DRVSR	25.35	0.7557	-	-	1) Can accept photos of any size as input. 2) Made use of an SPMC to control inter-frame motion more effectively. 3) Because the SPMC layer has no trainable characteristics, it can be utilized for testing with any scaling factors. 4) Following SPMC alignment, a DF network is used to successfully combine picture features from many images.

RRCN	32.35	0.9023	-	-	1) It was first suggested to use an extremely deep non-simultaneous completely RNN for VSR. 2) Multi-frame SR model and single-image SR model are combined using a model ensemble method. 3) Excellent representation capacity. 4) Capable of modeling both temporal and spatial non-linear mappings.
VSRResFeatGAN	25.51	0.7530	-	-	1) A generator up samples input frames, extracts feature as well as fuses them. 2) Discriminator evaluates the high-resolution frames quality produced
FFCVSR	26.97	0.8300	-	-	1) To concurrently store high-frequency characteristics and retain temporal consistency, outputs high-resolution preceding frames from misaligned low-resolution frames.
DUF	-	-	27.38	0.8329	1) uses deformable three-dimensional convolution to compensate for motion. 2) For specific input frames, the model estimates kernels.
FSTRN	-	-	29.95	0.8700	1)The rapid spatiotemporal residual block (FRB) simultaneously extracts spatiotemporal features while ensuring great computational efficiency. 2) The performance of feature extraction is improved by the residuals applied to the LR space. 3) Cross-space residual learning takes advantage of the similarities between the HR (High-Resolution) output and the LR (Low-Resolution) input.
DSMC	-	-	30.29	0.8381	1) In addition to extracting coarse spatial features and fine implicit MEMC, a U3D-RDN also reduces computational complexity. 2) MSCU module to aid in the use of intermediate data during upsampling. 3) A dual subnet that can improve generalization ability.

Table 3 summarises the effectiveness of a few VSR approaches based on both PSNR and SSIM as well as their benefits. Since Vimeo-90K is the largest dataset of VSR with actual scenes, it is the most often used training dataset. Vid4 has more high-frequency frames than any other dataset, making it the most used for testing. Thus, Vid4 is usually used to assess how well VSR works. Videos in REDS have a lot of movement, making VSR approaches difficult to use. System memory of more than 16GB required for implementing the algorithms. Coding is done in python by implementing modules like opencv, tensorflow, pytorch, cuda programming etc.

4. Conclusion

This survey examines the advancement of DL algorithms for video super-resolution. It has been discussed how to classify these algorithms and their designs. The effectiveness of various strategies on benchmark datasets is compared and analyzed, and the results are tabulated as well. The video frame resolution could be improved by the use of VSR techniques, which can also lead to improvements in visual quality and recognition precision. It could be utilized for a broad range of applications, like video decoding, remote sensing, 3D reconstruction, and medical diagnosis. One might choose DUF or DSMC for ultra-resolving videos with rich details and realistic textures but without jerky movements. As it provides a decent PSNR value for the REDS data set, DSMC can also be employed for complex and big motions.

5. References

[1] Yang W, Zhang X, Tian Y, Wang W, Xue J H and Liao Q 2019 *IEEE Trans Multimedia* **21** 3106

[2] Liu H, Ruan Z, Zhao P, Dong C, Shang F, Liu Y, Yang L and Timofte R 2022 *Artif Intell Rev* **55** 5981

[3] Bao W, Lai W S, Zhang X, Gao Z and Yang M H 2021 *IEEE Trans Pattern Anal Mach Intell* **43** 933

[4] Bare B, Yan B, Ma C and Li K 2019 *Neurocomputing* **367** 236

[5] Kappeler A, Yoo S, Dai Q and Katsaggelos A K 2016 *IEEE Trans Comput Imaging* **2** 109

[6] Li D, Liu Y and Wang Z 2018 *Proceedings - International Conference on Image Processing, ICIP* vol 2017-September p 17597474

[7] Shi W, Caballero J, Huszar F, Totz J, Aitken A P, Bishop R, Rueckert D and Wang Z 2016 *Proceedings of the IEEE Computer Society Conference on Computer Vision and Pattern Recognition* vol 2016-December p 16541213

[8] Tao X, Gao H, Liao R, Wang J and Jia J 2017 *Proceedings of the IEEE International Conference on Computer Vision* vol 2017-October p 17453269

[9] Lucas A, Lopez-Tapia S, Molina R and Katsaggelos A K 2019 *IEEE Transactions on Image Processing* **28** 3312

[10] Yan B, Lin C and Tan W 2019 *33rd AAAI Conference on Artificial Intelligence, AAAI 2019, 31st Innovative Applications of Artificial Intelligence Conference, IAAI 2019 and the 9th AAAI Symposium on Educational Advances in Artificial Intelligence, EAAI 2019* vol 33

[11] Jo Y, Oh S W, Kang J and Kim S J 2018 *Proceedings of the IEEE Computer Society Conference on Computer Vision and Pattern Recognition* p 18311890

[12] Li S, He F, Du B, Zhang L, Xu Y and Tao D 2019 *Proceedings of the IEEE Computer Society Conference on Computer Vision and Pattern Recognition* vol 2019 p 19263876

[13] Liu H, Zhao P, Ruan Z, Shang F and Liu Y 2021 *35th AAAI Conference on Artificial Intelligence, AAAI 2021* vol 3

[14] Ma Z, Liao R, Tao X, Xu L, Jia J and Wu E 2015 *Proceedings of the IEEE Computer Society Conference on Computer Vision and Pattern Recognition* vol 07

[15] Jin Y, Song H and Liu Q 2022 *Journal of Image and Graphics* **27** 295

Smart Greenhouse and Warehouse Monitoring with Disease Detection using Machine Learning

P B Savitha[1*], Ayushi Rai[1], Niharika Singh[1], Chinmayi C Keshav[1], and Neelambike VG[1]

[1] Department of Electrical and Electronics Engineering, Dayananda Sagar College of Engineering, Bengaluru, Karnataka, India

savithapb-eee@dayanandasagar.edu

Abstract. Global climatic changes in the recent era have resulted in a significant global food shortage. Where there is a vast population and a diverse agricultural sector, food security is a major problem. Growing competition for land, water, and energy will undermine our capital food production, and the effect of climate change will continue to be a challenge. The integrated greenhouse and warehouse provide quality and sustainable food production. In this paper an integrated green house with warehouse monitoring hardware implementation, with novel feature of customized disease detection with convolution neural network (CNN) algorithm is presented. The real time monitoring of data of different sensors are displayed on the screen/ dashboard developed by the android application. This aids farmers in making wise choices on crop management and disease detection of greenhouse crops and effective monitoring.

1. Introduction

According to the 2022 World Hunger Index, India ranks 107th position out of 123 countries. The Food and Agricultural Organization [FAO] in its report 'The State of Food Security and Nutrition in the World, 2021, estimates that around 189.2 million people are undernourished in India. By these statistics about 14% of the population in India is undernourished [1]. The traditional approaches for managing greenhouses frequently rely on labour-intensive, time-consuming, expensive, unreliable, and subject to human error. Manual monitoring is frequently used in conventional greenhouse management techniques, which results in delayed disease detection, ineffective treatment, and significant crop losses. A greenhouse offers a controlled environment that is suited to the needs of the vegetation being grown there. Microclimate and agronomic parameters have historically been recorded rather manually and unevenly. When compared to open field cultivation, artificially controlled greenhouses produce more crops per square meter because the microclimatic factors that affect crop yield are constantly monitored and managed to ensure the best possible environment is created [2]. Several countries heavily rely on greenhouse farming to cultivate healthy crops and keep them safe from pests and diseases. In order to forecast air temperature in agricultural greenhouses more accurately and to evaluate high and low temperatures quantitatively. Using a greenhouse model and MATLAB/Simulink, the researchers applied the suggested technique with success. With only roughly 1.5% of the projected temperature being off from the actual temperature, the model simulates and analyses climate controls accurately [3]. A GSM module controls automatic drip irrigation that is based on soil moisture criteria. Sensors control the temperature and humidity, and foggers aid in preserving the ideal humidity levels. A valve connects an above tank to the drip irrigation system, which fills based on predetermined criteria. Beehive boxes help with pollination and track how much honey is produced. For the access of an e-commerce corporation, data is kept in cloud storage containers. Ideal illumination and temperature conditions are maintained using foggers and LED lights. The Peltier effect, solar energy, and wind turbines all provide electricity for the greenhouse's machinery. In general, this strategy optimizes water use, improves pollination, and makes use of renewable energy sources for efficient and sustainable agriculture [4]. Hence, by combining IoT with embedded technology, many of the current practical issues have found solutions. The real-time data collection, analysis, and feedback from this system should allow for precise adjustments to be taken. To efficiently control moisture, temperature, and humidity inside the greenhouse by activating irrigation pipe, a cooling fan, and sliding windows in accordance with the conditions necessary for the crops to achieve maximum growth and yield [5]. The control system makes use of a motion sensor to activate an electric fence and a siren to ward off intruders or monkeys. Additionally, it has a function that disables the electric fence when no intruders are found [6]. There are obstacles to overcome while creating a green IoT-based agriculture solution, such as those related to hardware, data analytics, maintenance, mobility, and infrastructure. The hardware difficulties relate to the selection of sensors and range for IoT devices. As a result, there are many different types of sensors that can be employed in IoT applications (such as temperature, proximity, pressure, water quality, chemical, gas, and humidity sensors). Smart Warehouse that uses a variety of automated technologies and solutions to manage operations and are equipped with sensors and technologies. Need to create a comprehensive and a cutting-edge secured warehouse system to ensure the safety of inventory, data, and assets. In IoT-based monitoring system crops are stored and preserved in some prime state after harvest [7]. IoT sensors gather information on the warehouse climate and take prompt action in response to the farmers' input. Using controller and sensors, this system gives farmers access to real-time temperature and humidity data [6]. Individuals who are authorised to enter and exit by using the Radio Frequency Identification (RFID) tags are allowed. The farmer can use an android application to stay informed about the operation, be alerted to any crises, and receive alerts about parameter changes and actions made while continuously displaying the values of all the parameters [7]. The application of machine learning (such as deep learning methods) and prediction algorithms to IoT data is the key to solving the data analytics dilemma for smart agriculture [3]. An algorithm known as the Convolutional Neural Network (CNN) can identify different types of illnesses in leaves. This covers execution procedures such dataset collection, training, segmentation, feature extraction, testing, and classification. CNN is used to categorise the leaves as healthy or unhealthy based on the data. This project's implementation of real-time input from a dataset or Google source that was educated by the system aids in disease detection and offers solutions for overcoming deficiencies [8]. The disease prediction and detection aid in raising greenhouse output and profitability as a whole [7].

2. Proposed system

The proposed work is having three different sections namely green house, disease detection and ware house. The proposed greenhouse model block diagram is shown in figure 1 and ware house in figure 10. The shape of green house is designed with concept of traditional gable construction. The sensors used are temperature, IR (Infrared), rain, soil moisture, temperature and humidity, gas sensors for effective monitoring of greenhouse [8].

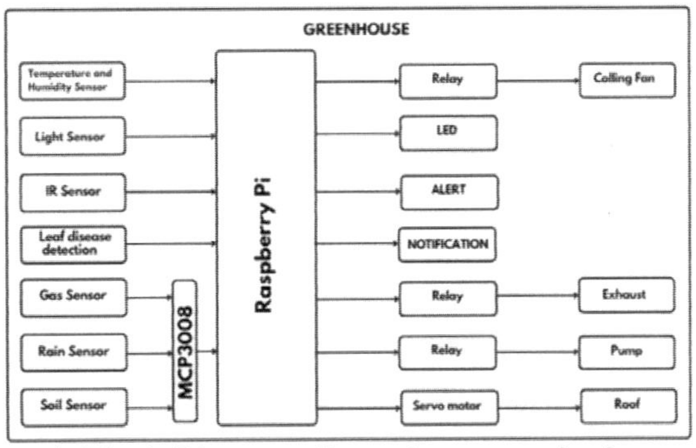

Figure 1. Block Diagram of greenhouse monitoring system.

The various sensor operations are tested for the prototype of dimension 40×40×50 cm. In the proposed greenhouse system, the IR sensor is placed at a distance of 10 cm apart and moisture sensor at 4 cm apart from each side of the wall and temperature sensor at the centre of the green house at 20 cm from either side of the wall. The gas and moisture sensors are placed at 10 cm above the wall. The monitoring of each sensor is explained with the general flow The green house is tested for chilly and tomato plant.

2.1. Temperature control system

Figure 2 shows the general glow chart of temperature sensor. The temperature is sensed with the help of DHT 11 sensor.

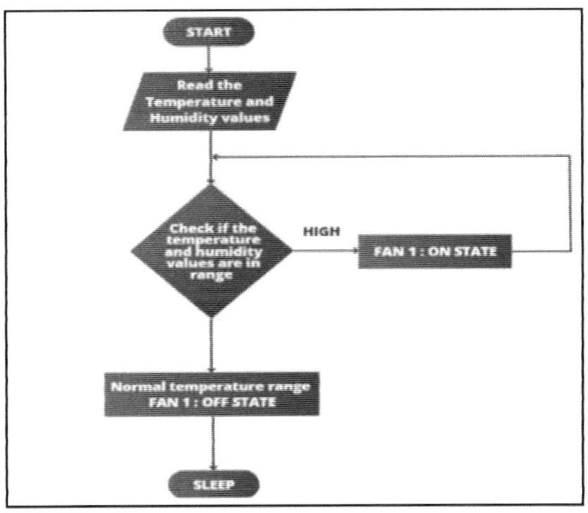

Figure 2. Flow chart of temperature control system.

2.2. Light control system

The integration of a Light-Dependent Resistor (LDR) module enables precise detection and regulation of light intensity. Figure 3 depict the Light intensity control system. This loop can be achieved using a while loop that repeatedly checks the light detection status and controls the LED accordingly.

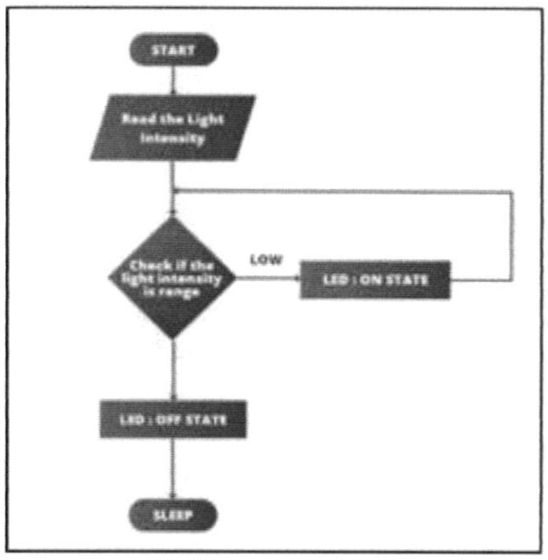

Figure 3. Flow chart of light control system.

2.3. Irrigation system

An irrigation system for a smart greenhouse can help automate the process of watering plants, ensuring they receive the right amount of water at the right time using soil moisture Sensor and rain sensor YL-63. Figure 4 shows thc irrigation system flow chart.

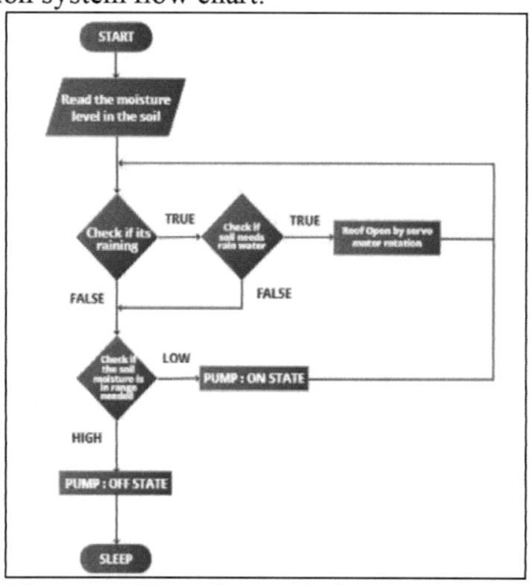

Figure 4. Flow chart of irrigation system.

When rain is detected, check the soil moisture level. Read the input from the soil moisture sensor and convert the analog value to an equivalent moisture level based on the sensor's calibration [9]. Compare the moisture level to a threshold value to determine whether it is low or not. If rain is detected and the soil moisture level is low, use the GPIO library to activate the GPIO pin associated with the relay module. This action turns on the power supply for the servo motor. The servo motor can then be used to open the roof, allowing rainwater to enter. Additionally, send an alert in the application indicating that it is raining. If rain is not detected but the soil moisture level is low, activate the GPIO pin connected to the relay module. This action turns on the power supply for the pump, which can be used to provide water to the soil. Also, send an alert in the application indicating low moisture levels. When the soil moisture level reaches the desired range, turn off the relay module to stop the pump and servo motor. This ensures that water is not continuously supplied when it is not needed.

2.4. Ventilation system

A ventilation system is an essential component of a smart greenhouse as it helps to regulate temperature, humidity, and air quality inside the greenhouse [10]. The MQ-135 sensor is used to identify various

IOP Conf. Series: Materials Science and Engineering 1295 (2023) 012010 doi:10.1088/1757-899X/1295/1/012010

gases in the atmosphere. The MQ135 sensor is capable of detecting NH3, NOx, alcohol, Benzene, smoke, CO2, and a few additional gases. Figure 5 displays the flow chart of ventilation system.

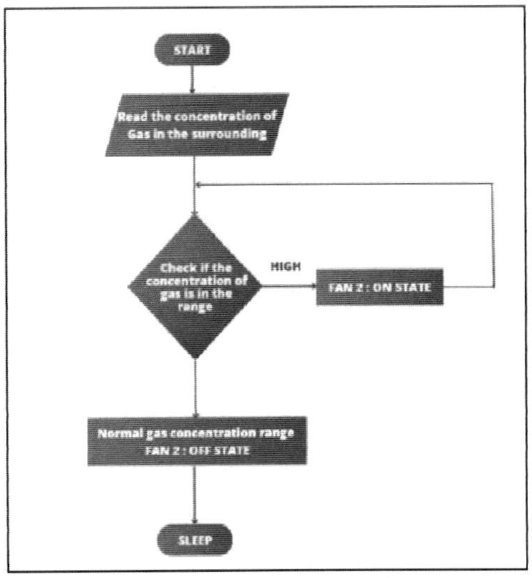

Figure 5. Flow chart of ventilation system.

If the gas concentration exceeds this threshold, use the GPIO library to activate the GPIO pin connected to the positive terminal of the DC fan. This action will provide power to the fan, enabling it to start and exhaust air from the greenhouse or enclosed area. Similarly, if the gas concentration falls below the threshold, turn off the GPIO pin connected to the fan, stopping the fan and conserving power [11].

2.5. Security system

The implementation of a security system inside a greenhouse holds significant importance in ensuring the protection, safety, and productivity of the greenhouse environment [12]. An IR sensor is used to detect the infrared radiations. Figure 6 shows the flow chart of Security System.

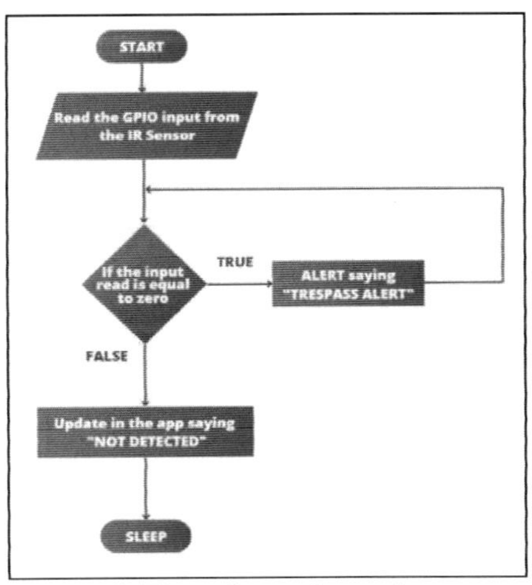

Figure 6. Flow chart of security system.

When an IR signal is detected on either of the GPIO pins, read the value of the corresponding input pin. If the value is equal to 0, it indicates that an IR signal has been received. In response to the IR signal, generate an alert notification. To keep track of the data and generate alerts, send the IR status information to the application on a regular basis. This ensures that alerts are generated promptly when an IR signal is detected.

IOP Conf. Series: Materials Science and Engineering 1295 (2023) 012010 doi:10.1088/1757-899X/1295/1/012010

2.6. Plant disease detection

Early detection of plant disease aids in the prevention of large-scale crop losses. Usually, to find and recognize diseases, farmers or other experts keep a constant eye on the plants. But this process is typically time-consuming, expensive, and inaccurate. Plant disease can be determined by searching for a spot on the infected plant's leaves that can help with detection. The objective of this study is to develop a disease recognition model that is backed by classification of leaf images. In this paper an image processing technique with a convolution neural network (CNN) is used to identify plant illnesses. Convolutional neural networks (CNNs) are useful for acquiring and analyzing data from images of leaves in order to determine healthy and diseased leaves of plants.

A convolutional neural network is network architecture for deep learning that learns directly from data. It provides an ascending approach to image classification, pattern identification, and object recognition tasks. A CNN has multiple layers that learn to detect different features of an image. The three main types of layers are the convolutional layer, the pooling layer, and the fully connected layer. The recognition and classification procedures are depicted in figure7.

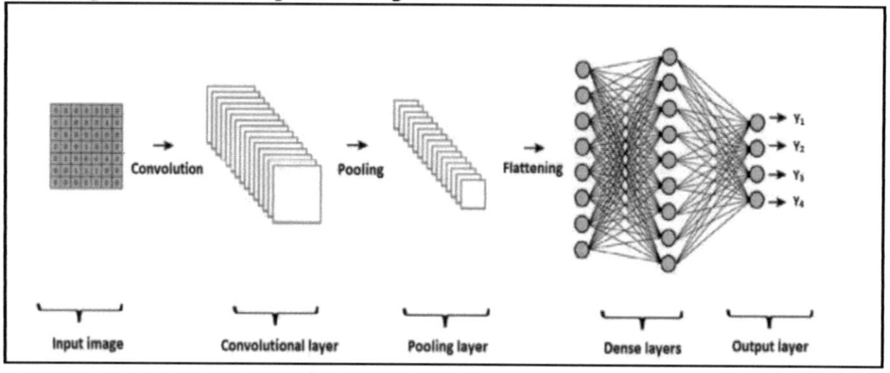

Figure 7. CNN architecture.

3. Plant disease detection using pi camera

The plant disease dataset contains images of plants in different classes. There are 10 different classes in this tomato crop dataset. Nine of them are diseased classes, and one is a healthy class. Data cleaning and augmentation techniques are performed on the images in order to enhance the performance of CNN models by eliminating blurry images and applying transformations such as rotation, flipping, or changing brightness and contrast [13]. CNNs learn to extract important features from the input image, calculate errors, update model parameters, and enhance the model's performance using back propagation. Once the model is trained, it is tested on a test dataset. The training algorithm is written to rescale, extract features and compare, and when trained for the desired images, it extracts features on multiple layers and gives out a hierarchical data format 5 file consisting of different models differentiated based on the features of the pictures and their resemblance with each other. This h5 file is used as a reference to compare the images that are passed during testing. Figure 8 displays the sample dataset of tomato plant disease.

Figure 8. Tomato Diseases - (a) Bacterial Spot, (b) Early Blight, (c) Late Blight, (d) Leaf Mold, (e) Septoria Leaf Spot, (f) Spider Mites, (g) Target Spot, (h) Tomato Mosaic Virus, (i) Tomato Yellow Leaf Curl Virus, (j) Healthy.

The Pi camera installed in the greenhouse, the images can be captured, it is pre-processed to resize and normalize the captured image. This image goes through the multiple layers of convolution and pooling of CNN after which it is predicted against the same model file obtained after training. The disease to which the test image has maximum similarities in terms of RGB values is displayed and updated. The RGB picture is converted into a grayscale image after which it is resized to match the other images used to train. It is then converted to a NumPy array and also normalized to reduce the complexity of CNN. Following it, is converted to 4D and sent to predict against the trained model. It returns an array whose values are scales of similarities between the captured image and each disease [14]. The index of maximum values is determined to know the disease predicted by the model by comparing the test image with each class. When a disease is detected, it is updated in the database which will dynamically update in the web application. An alert will be displayed on the app for the owner to take the required actions to minimize the breakout of the disease. This will predict the disease in the earlier stages so that suitable pesticides or measures can be taken so that the crops are not damaged.

3.1. Warehouse

A warehouse is a large, specially constructed building used for storing, organizing, and managing various kinds of goods, supplies, and inventory. Figure 9 shows the block diagram of the ware house system. The ESP32 acts as the central control unit, managing the connections and interactions between the RC522 RFID reader, IR sensors, flame sensor, motor, and the mobile application or central server. The system integration of these components allows for secure access control, real-time monitoring of space availability, and timely alerts in case of fire incidents.

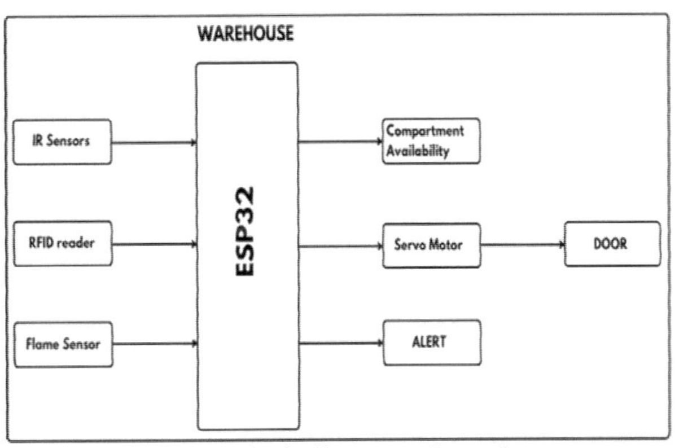

Figure 9. Block diagram of warehouse.

4. Hardware modelling and implementation

The proposed green house with warehouse and disease detection details is presented in the following section. The circuit diagram is presented in figure 9 and prototype model is in figure 10. The operation of the greenhouse model is explained with the help of flow charts and each sensor actions with relevant sensor images are displayed in table 1 The dash board of the application developed using MIT app inventor which enables to fetch data from the database and display them on one screen.

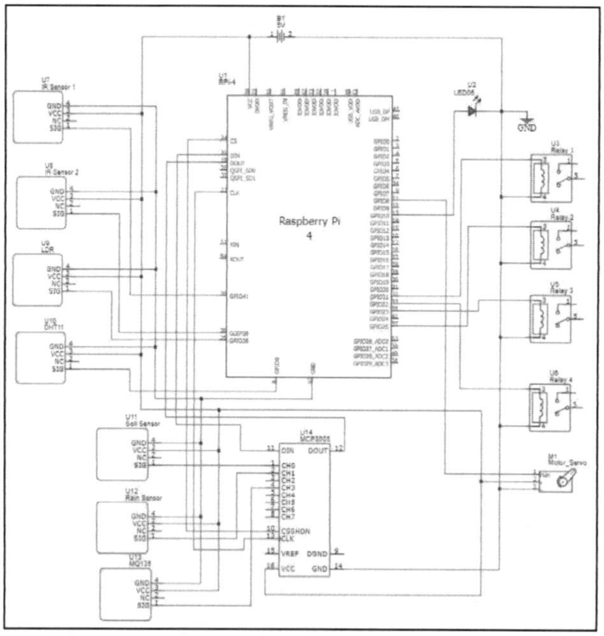

Figure 10. Connection diagram of greenhouse.

The Raspberry Pi micro controller and the 4-channel relay module is used to control high-voltage devices with low- voltage signals from the Raspberry Pi. Each channel on the relay module consists of a control input and a set of normally open (NO) and normally closed (NC) contacts. Connect the DC fan and DC pump to separate channels of the relay module. Each device should have a power supply (figure 11).

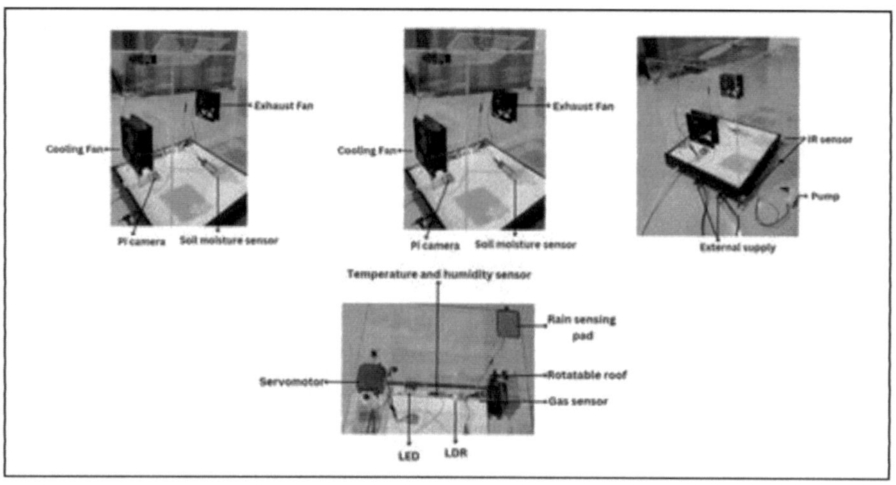

Figure 11. Hardware model of greenhouse system.

The climatic parameters that determine crop output are constantly monitored and adjusted to the optimum possible environment; artificial greenhouses yield more crops per square meter. This automated system keeps track of and regulates all environmental variables required for wholesome plant growth using microcontroller-controlled sensors and actuators.

Table 1. Specifications of greenhouse sensors.

Name of the sensor	Range chosen
Temperature (DHT11 Sensor)	24-27 °C
Soil Moisture (YL69 Sensor)	40–70%
Trespassing(IR Sensor)	2 – 10 cm
CO2 concentration in ppm(MQ135)	400 –1500 ppm
YL83 Rain Sensor	>30%

The connection diagram of ware house system is shown in figure 12. The system is controlled by ESP32 including an RC522 RFID reader, IR sensors, and a flame sensor, to provide access control and monitoring functionalities. The process begins with the RC522 RFID reader, which reads the unique identifier of RFID tags presented by users. This identifier is then compared to a list of authorized tags stored in the system's memory. If the tag is found to be authorized, access is granted. The system activates a motor that is connected to a mechanism responsible for opening the door, allowing authorized users to enter or exit the secured area. A notification is sent to a mobile application or central server to indicate whether access has been granted or denied. This provides real-time feedback to the user regarding their access status.

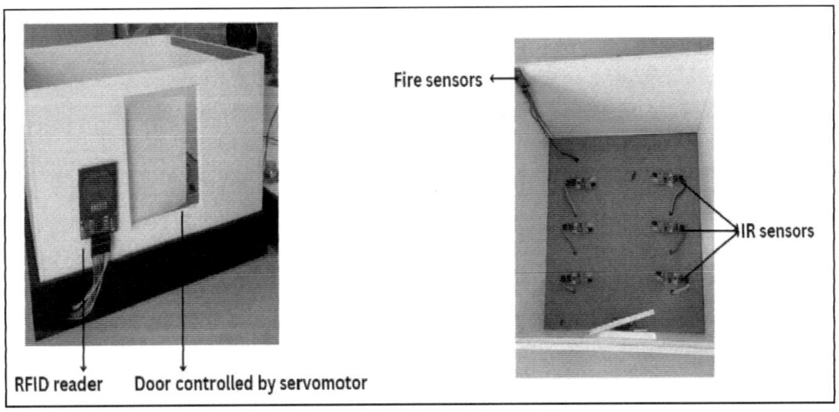

Figure 12. Hardware model of greenhouse system.

In addition to access control, the system includes six strategically placed IR sensors within a storage compartment. Figure 10 shows the IR Sensor Module. These sensors continuously monitor the presence of objects or obstacles within their respective areas. By analysing the status of the IR sensors, the system determines the availability of space in the compartment. The absence of objects detected by the sensors indicates vacant space. Based on the number of sensors detecting the absence of objects, the system calculates the percentage of available vacant space. This information is then communicated to the mobile application or central server, enabling users to remotely monitor the available space and make informed decisions about storage [15]. To ensure safety, the system incorporates a flame sensor that constantly scans the area for any signs of flames or fire. If flames are detected, the sensor triggers an alert mechanism within the system. This alert is sent as a notification to the mobile application, notifying the user of the potential fire or danger.

5. Results and discussion

The proposed system of smart green house, ware house and disease detection feature with CNN algorithm is developed and hardware is implemented. Hardware model is tested for temperature, rain fall, soil moisture; tress pass. The use of Raspberry Pi increases the accuracy in terms of time and action against detection of each parameter based on the predetermined condition. The results are shown in table 2 for green house monitoring and table 3 for ware house monitoring. The web application using MIT app inventor results are displayed on the dashboard and is shown in figure 13. The web application displays all the parameter values that are dynamically updated. The Table 3 shows the disease detection using CNN algorithm is presented. The accuracy of the algorithm for a data set of 1000 is tested and the result shows effective detection of the disease. Diseases that have been detected alerts the user for the required condition or situation.

Table 2. Result of greenhouse monitoring system.

Name of the sensor	Reason
Access (RFID Tag)	Right Card
Access (RFID Tag)	Wrong Card
Fire Sensor	Low Signal on the Output
Storage (IR Sensor)	100% Empty, 50% empty and 83.33% empty

Table 3. Healthy and diseased plant leaves of tomato and chilly samples.

Type of sample leaves status	Test Image	Predicted with CNN algorithm
Tomato – Healthy		Healthy
Tomato – Late Blight Disease		Tomato – Late Blight Disease
Chili -Healthy		Healthy
Chili – WhiteflyDisease		Chili – Whitefly Disease

6. Conclusion and future scope

In this paper, the hardware implementation of integrated smart green house with warehouse and an additional feature of disease detections is presented. The proposed system is designed and tested for the tomato and chilly plant. The various sensors used are temperature, humidity, soil, moisture, and IR sensor. The proposed unit is monitored remotely with the help of web application using MIT app inventor. The disease detection technique of plant is realized with CNN algorithm. The proposed work helps in early disease detection, improved crop health and yield, alerts and real time monitoring and data driven decision. The hardware is designed and implemented to monitor greenhouse effectively. Future scope of the proposed work may be by integrating block chain technology to the ware house monitoring unit.

7. References

[1] Lodhi S, Sakshi and Kukreja V 2022 *2022 International Conference on Data Analytics for Business and Industry, ICDABI* p 22627786
[2] Danita M, Mathew B, Shereen N, Sharon N and Paul J J 2019 *Proceedings of the 2nd International Conference on Intelligent Computing and Control Systems, ICICCS* p 18510966
[3] Khan F A, Ibrahim A A and Zeki A M 2020 *J Phys Commun* **4** 5
[4] Jiang S, Zhang M, Li X, Qi Y and Lü X 2022 *Journal of Chinese Agricultural Mechanization* **43** 159
[5] Saleem M H, Potgieter J and Arif K M 2019 *Plants* **8** 468
[6] Anoop A, Thomas M and Sachin K 2021 *2021 Asian Conference on Innovation in Technology, ASIANCON* p 21224883
[7] Suresh S and Chakaravarthi G 2022 *Measurement (Lond)* **195** 111197
[8] Gong L, Yu M and Kollias S 2023 *Algorithms* **16** 243
[9] Rayhana R, Xiao G and Liu Z 2020 *IEEE Journal of Radio Frequency Identification* **4** 195
[10] Fisher D K and Kebede H 2010 *Comput Electron Agric* **74** 168
[11] Liu Z, Bashir R N, Iqbal S, Shahid M M A, Tausif M and Umer Q 2022 *IEEE Access* **10** 44934
[12] Vishnoi V K, Kumar K, Kumar B, Mohan S and Khan A A 2023 *IEEE Access* **11** 6594
[13] Deshwal P, Sharma K and Moudgil Mr S 2023 *Int J Res Appl Sci Eng Technol* **11** 6594
[14] Hamad I H, Chouchaine A and Bouzaouache H 2021 *18th IEEE International Multi-Conference on Systems, Signals and Devices, SSD* p 20747476
[15] Harakannanavar S S, Rudagi J M, Puranikmath V I, Siddiqua A and Pramodhini R 2022 *Global Transitions Proceedings* **3** 305

Thermal Power Plant Pollution Monitoring and Control using Internet of Things

Mahalakshmi Manoharan[1*], Padma S[1], Madhubalan S[1], Arunkumar S[1], and Kesavan D[1]

[1] Department of Electrical and Electronics Engineering, Sona College of Technology, Salem, India

mahaeee56@gmail.com

Abstract. In our day-to-day life, environmental degradation and global warming are essential factors to consider. These problems were caused due to the contamination of water, air, and soil with the waste generated by industries. The pollutants level generated by the industries to be measured and controlled by central and state pollution control board. The air pollutants are the major cause for the deterioration. The existing pollutant monitoring system are complicated, inaccurate, unreliable, delayed intimation and less efficient. Hence in mettur thermal power plant, the internet of things-based pollutant gas monitoring system has been implemented. This system consists of numerous gas sensors, Arduino UNO controller, ESP8266 Wi-Fi module, ubidots server and alarm. The pollutants measured are carbon monoxide (CO), suspended particulate matters (SPM), sulphur oxides (Sox), nitrogen oxides (Nox) in air. When the pollutants level of industry is crossed the limit, the pollution control board has been intimated the exact location of the industry through internet. The pollution control board can automatically send alert SMS and email to the respective industry. This system is suitable for the recent technology industrial internet of things. The electrostatic precipitator has been installed for pollutant control. It is proved that the proposed system is a very good replacement for the conventional monitoring system done by the human. It is acted as user friendly interface for the operation of the system and is reliable. The system has attained improved pollutant control, increased stability, and consistency of the power plant and finally the safety of the employees working in the power plants and mine areas.

1. Introduction

Human health and urban air pollution has been closely related [1]. Because of the incredibly quick expansion of industrial development and urbanization, the natural air environment has been contaminated and is rapidly degrading. Therefore, it is crucial to monitor and limit such pollutants if we want to avoid environmental catastrophes. Conventional analytical tools are time-consuming, expensive, and rarely employed in the field in real-time [2]. Research has been proved that the atmospheric pollutants are the major determinant factor which caused the chronic respiratory disease obstacle in human beings. According to the specialized cancer agency of world health organization, the International Agency for Research on Cancer (IARC), it proclaimed that outdoor air pollution is causing the carcinogenic to humans and lung cancer [3]. These pollutants lead to huge economic losses and affects environmental development. The pollutants are generated due to air, water and soil contamination with the waste emitted from industries. The outcome of this is environmental degradation. Technological development cannot be omitted due to the increased population growth, leading to more pollutants. The industrial pollution issue has taken as grave importance for agencies trying to fight against environmental degradation. All around the world consider it a serious issue and the necessary actions to be taken.

In our state, the industries are directly emitting pollutants through their chimneys to the sky. The upshot of this is environmental deterioration. One third of atmospheric pollution is due to the industries pollutants. The environmental protection department facing some difficulties in monitoring the industrial waste emission. To help them, the remote gas monitor system has been executed to measure the pollutants coming out from the chimney of large industries. Using the gas sensors, this system measured the emitted pollutants such as sulphur oxides, carbon oxides, and nitrogen oxides and transferred these data to the central server via controller and RS232. With the help of data received the exact chimney position can be detected and monitored remotely by environmental protection department. It has main contribution of this paper is the polluted air emitted from the chimney has been detected by the IoT based wireless gas monitoring system. This is communicated to the pollution control board through internet with the exact location of the industry. The PCB can take necessary actions if the pollutant levels are crossed their tolerance value in the air. The alert SMS and email is sent to the industry. This system designed to attain the safety of human beings working in power plant and mines, to measure the pollutants level remotely and to control the air pollution level thereby reduced the environmental deterioration.

2. Literature survey

Nowadays, the government has measured the industrial pollutants manually which is not reliable. Hence the people did not have proper awareness about the pollution control. The country needs industrial pollution monitoring and analysis. Some of the literature papers has been reviewed here. The World Health Organization (WHO) has explained the air quality and its rationale, its role for protecting human health, particulate matter, ozone, nitrogen dioxide, sulphur dioxide and the risk assessment summary [1]. Air quality assessment has to be made to protect the human beings from toxic air pollutants. The assessment report helps calculate the air quality and human impact assessment. Human exposure to the air pollution can affect the human health in indoor and outdoor areas. In this, the designing and planning of the monitoring system, cost effectiveness, quality assurance and control and managing and disseminating information to human beings are discussed [4]. In 1998, accordant with the clinical data and environmental monitoring stations in Taiwan, the author estimated the relationship between air pollution and daily number of patients coming to clinic due to lower respiratory tract illness. The clinic visits are based on the concentrations of nitrogen dioxide, carbon monoxide, sulphur dioxide and other pollutants in air on daily basis. This analysis uses the small area design and hierarchical modelling [5]. The United states Environmental Protection Agency (EPA) designed a miniature model for next generation air monitoring, which forecasted the support for successful installation and implementation of new monitoring technologies. It acted as a sample framework for organizations and agencies with same type of goals, coordination, and collaboration [6]. The wireless sensor network application for environmental monitoring has been implemented for the real time monitoring of unprotected agriculture and habitat, indoor living monitoring, greenhouse monitoring, climate monitoring and forest monitoring [7]. For monitoring and controlling highly variable emission sources, such as in metropolitan regions, spatial and temporal scales measurements at pertinent levels are crucial. However, because traditional air quality measurement techniques are expensive and complicated, In general, measuring networks are very sparse. In this study, we demonstrate that inexpensive, electrochemical gas sensors have been employed to detect parts per million (ppm) mixing ratios be implemented for parts-per-billion (ppb) level research for relevant gases when properly designed and operated to the smog in cities. At fine spatial and temporal scales, the low cost scalable high-density air quality sensor networks for static and mobile configurations is made possible by great portability of sensor nodes using lot of separate electrochemical sensors. The electrochemical sensors performance is demonstrated in this paper, and the results of deployments of sensor node networks in autonomous, high-density, static networks as well as mobile networks for measuring personal exposure are then described. In urban areas, CO, NO and

NO2 electrochemical gas sensors are utilised to measure the high granularity gas quality structure, monitoring and implemented for human exposure quantification [8]. An online GPRS sensors array monitoring has been depicted which contains a fixed internet enabled pollution monitoring server and mobile data acquisition unit (DAQ). The mobile DAQ collects the air pollutants level of the gases CO, NO2, and SO2 and pack them with GPS physical location, time and date and is uploaded to GPRS modem and transferred to server through public mobile network. The system is implemented in Sharjah, UAE used to measure and report the real time pollutants level and their location on 24/7 basis to the server [9].

Scikit-learn is a python module used to integrate the numerous machine learning algorithms for medium learning supervised and unsupervised problems. It provided the ease of use, improved performance, documentation, API consistency, minimal dependencies and can be used for academic and commercial settings. Since it depends on the scientific Python ecosystem, it is applicable to all types of statistical data analysis. It's significant because the algorithms, which are implemented in a high-level language, can serve as the foundation for strategies particular to a use case, like medical imaging. It can be used as online learning tool for large datasets [10]. Irrespective of the learning sample output values, the new tree-based ensemble method for supervised and regression problems is depicted to build the randomized trees in the extreme cases. The computational efficiency is increased [11]. Random forests is an effective tool for prediction and by providing the right kind of randomness the accurate classifiers and regressors can be attained with lower error rates. By emphasizing the potency of the individual predictors and their connections, the framework also sheds light on the random forest's predictive capability. By using an out-of-bag estimate, theoretical strength and correlation values are given a concrete meaning. The use of arcing and boosting techniques can reduce bias and variance. In order to reduce bias, regression's adaptive bagging method which also works well in classification and classification was developed. However, just like arcing, it also modifies the training set as it goes along. Another way to think of random forests is as a Bayesian method. In classification but not regression, random features and random inputs provide good results. Bagging and random features are the only types of randomizations used in this investigation [12].

The hybrid model contained both autoregressive integrated moving average (ARIMA) and artificial neural networks (ANNs) to improve time series analysis forecasting performance in both linear and nonlinear models. The forecasting performance can be boosted using this complication method for complicated issues with both linear and nonlinear correlation patterns. The empirical findings using three actual data sets in which the hybrid model can perform well when considering each component model separately. Furthermore, utilizing the hybrid approach helps lessen model uncertainty often present in statistical inference and time series forecasting due to the potential for unstable or shifting patterns in the data. Also, the overfitting issue more closely associated with neural network models can be mitigated by first fitting the ARIMA model to the data [13]. The status of IOT development in China has been discussed in the following perspectives: policies, research and development plans, applications, and standardization to meet the architecture challenge. It discussed the opportunity and prospect of IOT [14]. Utilizing sensors, connection, embedded software, and Internet of Things, indoor air quality may be monitored in real time. The Web of Things is one strategy that is gaining ground that employs web architecture and web standards as a platform for developing IoT applications. An IoT platform for tracking indoor air quality is presented in this study [15]. The sensor devices that the concept was based on could sense, compute, and communicate data in a network. This study monitored the Air Quality Index, temperature, humidity, and PM2.5 in real-time. Monitored data is wirelessly transmitted via Wi-Fi module to a server. The public must get the awareness regarding the state of air pollution and its importance [16]. By using the Nova PM Sensor SDS011 and a BME280 sensor to detect temperature, pressure, humidity, and altitude, the real-time air quality monitoring system is used to measure PM2.5 and PM10, invisible particles that move in the air [17]. IOT Base Air Pollution Monitoring System uses an Internet-connected web server to track the air quality. When the number of dangerous gases, such as CO2, smoking, alcohol, benzene, NH3, and NOx present in the air exceeds a particular threshold, the device will sound an alarm [18].

The node MCU related programming are explained in this paper [19]. To assess the depicted model performance the error metrics values RMSE and MAE, which one is better has been discussed in this paper [20]. The recent success when implementing machine learning in real world problems and five basic learning paradigms are discussed [21].

3. Methodology

Across the country, the national air quality monitoring programmed (NAMP) operating stations are used to monitor the pollutants generated by the industries and it is under the control of central and state pollution control boards. Each industry is emitting the pollutants in different forms such as solid, liquid and gas, etc. these data are individually measured and stored in data logger using various methods. Tamil Nadu pollution control board (TNPCB) is monitoring all industries' air, noise, and water pollution individually throughout Tamil Nadu. It is taking control actions against industries who didn't follow the standards such as issuing notices, legal actions, closure order, water supply and power supply cut etc.It is safely disposing the industrial hazardous wastes and creating awareness between the people through the Environmental Training Institute, Environmental Awareness Cell, and conducting Environmental Awareness Programmes, etc.
The mettur coal based thermal power station (MTPS) is in Salem district is generating total power of 420MW. The existing pollutant measurement system provide unreliable data, inaccurate data, lag of data communication, and data may loss. Hence the proposed system is depicted for remote measurement of pollutants using IoT without any time delay. This system is increased the efficiency and accuracy which helps the central and state pollution control boards for monitoring and maintaining the environmental pollution.
The four different types of analysers are used to measure various gases as carbon monoxide (CO), suspended particulate matters (SPM), sulphur oxides (Sox), nitrogen oxides (Nox) in air. And these data are transmitted from MTPS to pollution department using RS485 communication.

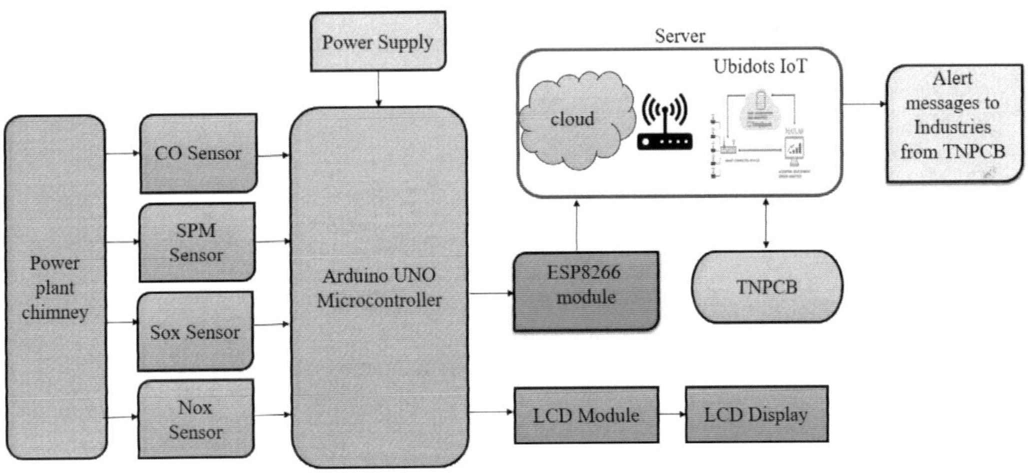

Figure 1. Block diagram of the proposed system.

The pollution department transfers these data to TNPCB through Wi-Fi broadband. The TNPCB can monitor and do analysis from the received data and data can be stored for future analysis too. The TNPCB can create awareness to the people based on the environmental pollutants level. The block diagram of the wireless gas monitoring system has been drawn in figure 1. The system consists of four different gas sensors which is taking input from the chimney of power plants and giving the emission level as input to the Arduino UNO controller. All the data are transferred to TNPCB central cloud server through EP8266 Wi-Fi module. When the tolerance limit is crossed then it is indicated to TNPCB and displayed in LCD. Then the TNPCB can take necessary actions if required. The pollutant analysis has been made as followed. In India, coals used for power plants are generally of high ash content as 30 to

50%.and contains sulphur up to 1%. He fly ash is the major pollutant followed by SO_2, CO and NOx. The annual coal consumption for the four plants totally is as followed.
Coal consumption 137 Tph. Per unit. Coal consumption for 4 units = 137 * 4 = 584Tph.Coal consumption for 4 units = 584*24*365 =5115840 Tones
The coal analysis for thermal power plant, calculated combustion products from burned coal and the emission rates for the gases emitted are mentioned in table 1, table 2 and table 3 respectively.

Table 1. Coal analysis for MTPS.

Parameters	Mean level
Hydrogen	2.74%
Carbon	37.73%
Nitrogen	0.74%
Sulphur	0.44%
Ash	44.84%
Moisture	6.17%
Oxygen	7.34%
Gross calorific value	3696 k Cal/kg

Table 2. Calculated combustion products from burned coal.

Combustion product	Total emission rule
SO2	3973 kg/h
NOx	254.1 kg/hr
H2O	9.22%
02	23.84%
Gross calorific value	3698 k Cal/kg

Table 3. Emission rates for each pollutant emitted.

Emission product	Emission rate
SPM	591.31 kg/hr
SO2	3939.84 kg/hr
NOx	249.48 kg/hr
CO	13.176 kg/hr

The MTPS implemented the electrostatic precipitator (ESP) to observe the dust and improve the system's efficiency. The ESP collects the 99.9% or more of the dust from the gas stream. Hence the proposed system improved the system's overall performance and environmental pollution has been controlled.

4. Result and discussion

The hardware image is mentioned figure 2. The hardware consists of oxygen sensor, carbon monoxide sensor, particulate emission sensor, NOx sensor, gas sensor, Arduino UNO controller, ESP8266 module, power supply module, LCD module and step-down transformer. With the help of power supply module suitable power supply is given to the circuit. The results are taken from ubidots server page. When the gas values cross the limit, it is indicated to the TNPCB, and the alert message and email are automatically sent to the respective industry. The outputs are shown in figure 3.

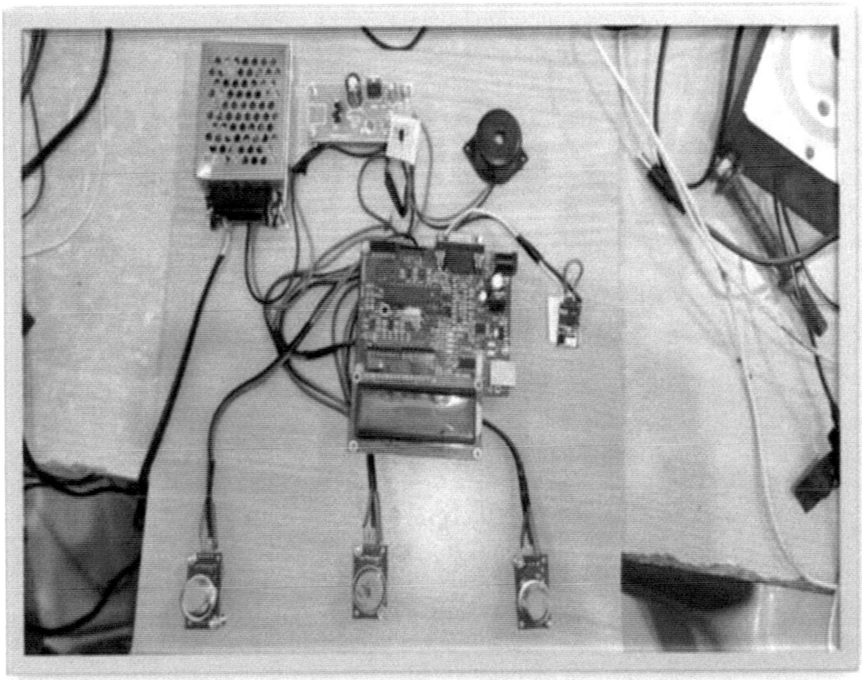

Figure 2. Hardware image of gas monitoring system.

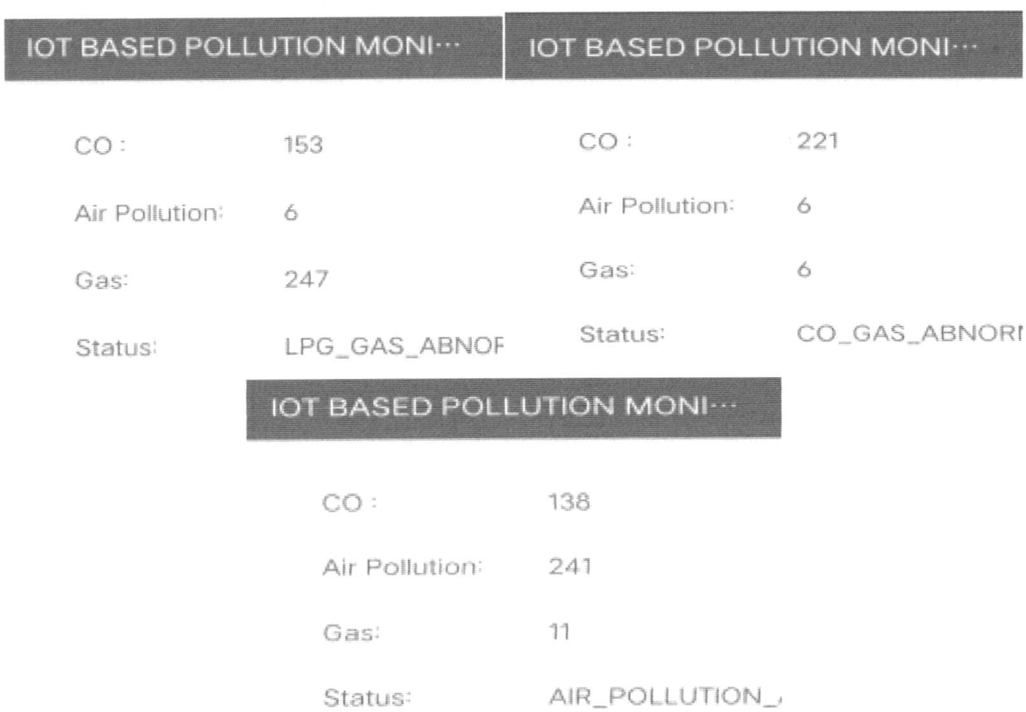

Figure 3. Output indication.

5. Conclusion

Nowadays pollution is the major problem to be considered to live healthy life. Industrial pollution is one of the most important pollutions to be considered and controlled which is done by central and state pollution control board. The case study has been done in mettur thermal power plant regarding the pollutants generated and their methods to control the pollutants level. Five different types of analysers are used to monitor the pollutants emitted. The existing system in the site is, the sensor values are transmitted remotely using RS485 to the pollution department. Then using broadband, the pollutant values are transmitted to Tamil Nadu pollution control board. This leads to some data loss, time lag and give unreliable data. Thus, these drawbacks have been overcome by designing IoT based gas monitoring system. Thus, the time lag, data loss problems are eliminated. By using IoT, the pollutants of multiple industries can be collected at a time by pollution control board. From the data collected the analysis can be done by TNPCB. When the pollutant values exhausted the tolerance limit, then it is intimating the exact location of the industry to TNPCB. The automatically generated alert SMS and email is sent to the industry. This technology is more suitable for the upcoming technology industrial internet of things (IIOT 4.0). This system is more reliable, efficient and have better future scope for environmental pollution monitoring and controlling.

6. References

[1] World Health Organization and Organization W H O W H 2005 WHO Air quality guidelines for particulate matter, ozone, nitrogen dioxide and sulfur dioxide *global update 2005. WHO/SDE/PHE/OEH/* **0602**

[2] Lee D D 2001 Environmental gas sensors *IEEE Sens J* **1** 214

[3] Xie S H, Liu A L, Chen Y Y, Zhang L, Zhang H J, Jin B X, Lu W H, Li X Y and Lu W Q 2010 DNA damage and oxidative stress in human liver cell L-02 caused by surface water extracts during drinking water treatment in a waterworks in China *Environ Mol Mutagen* **51** 229

[4] Bower J, Hanninen O, Kotlik B, Mucke H G, Ozkaynak H, Tarkowski S, Krzyzanowski M, Baumann R, Bozo L, Chicherin S S, Economopoulos A, Filliger P, Guler C, Hamekoski K, Larssen S, De Leeuw F, Lindskog A, Ruuskanen J, Sabiliauskas R, Santroch J, Schwela D, Slotova K, Stedman J, Wilson W E and Tonnisen K 1999 Monitoring ambient air quality for health impact assessment *World Health Organization Regional Publications - European Series* 1999 p 85

[5] Hwang J S and Chan G C 2002 Effects of air pollution on daily clinic visits for lower respiratory tract illness *Am J Epidemiol* **155** 1

[6] Zhang P, Shi X, Khan S U, Ferreira B, Portela B, Oliveira T, Borges G, Domingos H, Leitão J, Mohottige I P, Gharakheili H H, Moors T, Sivaraman V, Najari N, Berlemont S, Lefebvre G, Duffner S, Garcia C, Parmentier A, Cohen R, Nemani P, Krishna G S, Ramisetty N, Sai B D S, Kumar S, Constantinou G, You S, Shahabi C, Taesiri M R, Macklon F, Bezemer C-P, Roychowdhury S, Professor A, Rashid M T, Wang D, Sun N, Zhang J J, Gao S, Zhang L Y, Camtepe S, Xiang Y, King I J, Huang H-W H, Pierce L, Khairunisak, Kusumasari T F, Fauzi R, Zhu L, Xu M, Zhou A, Qiao J, Rao J, Ao T, Dai K, Zou X, Gao Y, Gao X, Yang X, Liu J-K, Chen G, Chen D, Tong D, Yang C, Cheng X, Shiina R, Tamaki S, Huang C-C, Hatano T, Yamada T, Shimada T, Taniguchi T, Du J, Lu Y-H, Li C-Y, Huo X, Ding Y, Li X, Huang J-L, Yin W, Dong J, Ji Y, Xiao Z, Hu X, Xiao J, Zhang G, Zhou Q, Li H, Cheng P, Chen Z, Xiao N, Peng Z, Feng X, Liu M, Yang Y, Su H, Xie H, Liang Y-P, Li Y, et al 2019 IEEE Draft Standard for Spectrum Characterization and Occupancy Sensing *IEEE Access* **9** 1

[7] Othman M F and Shazali K 2012 Wireless sensor network applications: A study in environment monitoring system *Procedia Engineering* **41** 1204

[8] Mead M I, Popoola O A M, Stewart G B, Landshoff P, Calleja M, Hayes M, Baldovi J J, McLeod M W, Hodgson T F, Dicks J, Lewis A, Cohen J, Baron R, Saffell J R and Jones R L 2013 The use of electrochemical sensors for monitoring urban air quality in low-cost, high-density networks *Atmos Environ* **70** 186

[9] Al-Ali A R, Zualkernan I and Aloul F 2010 A mobile GPRS-sensors array for air pollution monitoring *IEEE Sens J* **10**
[10] Barupal D K and Fiehn O 2019 Generating the blood exposome database using a comprehensive text mining and database fusion approach *Environ Health Perspect* **127**
[11] Aminifar A, Shokri M, Rabbi F, Pun V K I and Lamo Y 2022 Extremely Randomized Trees with Privacy Preservation for Distributed Structured Health Data *IEEE Access* **10**
[12] Jin Z, Shang J, Zhu Q, Ling C, Xie W and Qiang B 2020 RFRSF: Employee Turnover Prediction Based on Random Forests and Survival Analysis *Lecture Notes in Computer Science (including subseries Lecture Notes in Artificial Intelligence and Lecture Notes in Bioinformatics)* vol 12343 LNCS
[13] Zhang P G 2003 Time series forecasting using a hybrid ARIMA and neural network model *Neurocomputing* **50**
[14] Chen S, Xu H, Liu D, Hu B and Wang H 2014 A vision of IoT: Applications, challenges, and opportunities with China Perspective *IEEE Internet Things J* **1**
[15] Esquiagola J, Manini M, Aikawa A, Yoshioka L and Zuffo M 2018 Monitoring Indoor Air Quality by using IoT Technology *Proceedings of the 2018 IEEE 25th International Conference on Electronics, Electrical Engineering and Computing, INTERCON 2018*
[16] Landge P D and Harne R R 2018 Air Quality Monitoring System for City : A Review *International Research Journal of Engineering and Technology (IRJET)* **5**
[17] Truong T P, Nguyen D T and Truong P V 2021 Design and deployment of an IoT-based air quality monitoring system *International Journal of Environmental Science and Development* **12**
[18] Kumar R, Singla S and Bansal J 2023 IOT Based Air Pollution Monitoring System *International Scientific Journal of Engineering and Management* **02**
[19] Azrol Bin Azahar M, Shamian Zainal M, Shahidah Mohd Shah N, Md Nor D, Anuar Hamzah S and Mohamed M 2020 IoT-Based Air Quality Device for Smart Pollution Monitoring *Progress in Engineering Application and Technology* **1**
[20] Chai T and Draxler R R 2014 Root mean square error (RMSE) or mean absolute error (MAE)? -Arguments against avoiding RMSE in the literature *Geosci Model Dev* **7**
[21] Langley P and Simon H A 1995 Applications of Machine Learning and Rule Induction *Commun ACM* **38**

Design and Implementation of IoT based PV Integrated Power Monitoring System

Brinda S[1*], Rajdeep Singh[1], P B Savitha[1], Raunak Singh[1], and Tamseel Ahmad[1]

[1] Department of Electrical and Electronics Engineering, Dayananda Sagar College of Engineering, Bengaluru, Karnataka, India

brinda.anitha@gmail.com

Abstract. The awareness of usage of power and effective utilization of power is based on intelligent decision, which can help the user as well as supplier. In this paper the design and implementation of IOT based PV integrated system is presented. A hardware prototype is developed with PV module connected to boost converter, Arduino Uno controller, ESP 8266 Wi-Fi module, and Thing speak to monitor a real-time power. The Arduino Uno serves as the central controller, gathering data from smart power meters and PV modules for solar energy generation. The collected data is wirelessly delivered to the ThingSpeak, which provides cloud-based storage, data analysis, and visualization. With these users may remotely monitor usage of power, track solar energy generation, and get notifications for energy utilization.

1. Introduction

The fast growth of Internet of Things (IoT) has opened the way for cutting-edge solutions in a variety of fields, including power monitoring and management. With increased concerns about energy efficiency, cost savings, and sustainability, intelligent systems that can monitor and optimize power use in real-time are in high demand. IoT with cloud-based monitoring has been marked as a versatile tool for visualization and compatibility with other devices and transmit to cloud by making an idea for low-cost solutions. In answer to this demand, this paper provides an IoT-based smart power monitoring system. The goal of this paper is to provide an intelligent and user-friendly power monitoring system that enables people to make educated decisions about their energy consumption. The system provides power usage patterns, recognizing possible energy waste, and offering personalized optimization recommendations. Furthermore, the incorporation of renewable energy sources via the PV module adds to the use of clean and sustainable energy.

2. Literature survey

The integration of IoT-based technologies has significantly transformed energy management and control paradigms. These advancements encompass various aspects, including IoT-driven smart energy meters that facilitate the cloud-based analysis of consumption patterns [1]. The SMACS system employs IoT elements to monitor power usage and leverage cloud platforms for comprehensive data analysis [2]. These initiatives collectively bolster energy management both at the grid level and in everyday living scenarios, addressing critical challenges while enabling real-time measurements for accurate load analysis [3]. Expanding on this, the impact of IoT extends to seamless remote management and monitoring processes [4]. Moreover, the integration of IoT applications within the Smart Grid framework offers multifaceted benefits. Demand-side energy management, exemplified by systems like ISEMS, leverages sensor-based data collection and cloud-powered analysis to harness renewable energy and optimize usage of power [5]. Furthermore, IoT's transformative influence extends to areas like solar energy enhancement, where IoT technologies enable real-time tracking of factors like dust accumulation on solar panels, ultimately optimizing power production [6]. Venturing into novel territories, the field witnesses groundbreaking advancements. For instance, the adoption of a unified model predictive control approach in integrated PV and battery storage systems promises higher reliability and compliance with standards [7]. Simultaneously, the digital implementation of control loops for bidirectional converters, enriched with features like maximum power point tracking (MPPT) and high-bandwidth current regulation, introduces improved performance and enhanced energy flow management [8]. These innovative strides significantly contribute to the efficiency of renewable energy systems. However, amidst these promising advancements, challenges remain, particularly in dealing with higher voltage levels in PV and battery systems [9]. Despite the potential gains, the complexities of system design, installation intricacies, and safety concerns associated with elevated voltage levels necessitate careful consideration. This confluence of innovative solutions and persistent challenges drives the continuous evolution of energy management and control within the broader context of IoT-driven advancements.

3. Methodology

The proposed system is designed and implemented with the help of block diagram shown in figure 1. The PV panel output is converted to AC using H bridge inverter and is fed to a load. The switching signals for using boost converter and inverter are provided through Arduino Uno controller to driver IC. The maximum power point tracker is implemented with perturb and observe algorithm and is realized through boost converter. At PV side and load side voltage and current sensors are connected. The real time data is fetched to cloud via ThingSpeak APP and the flow chart is shown in figure 2.

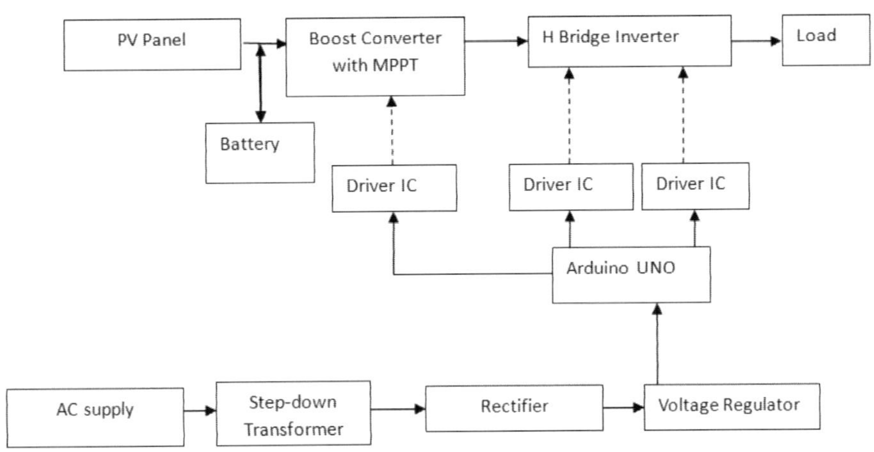

Figure 1. Block diagram of proposed PV based power monitoring system.

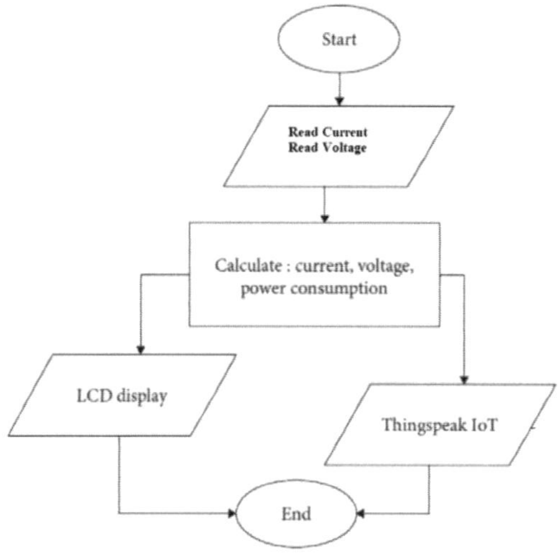

Figure 2. Flow chart of PV based power monitoring system.

4. Design

The PV power system is simulated using MATLAB/Simulink platform. The Simulink diagram is shown in figure 3. The designed values are mentioned in table 1.

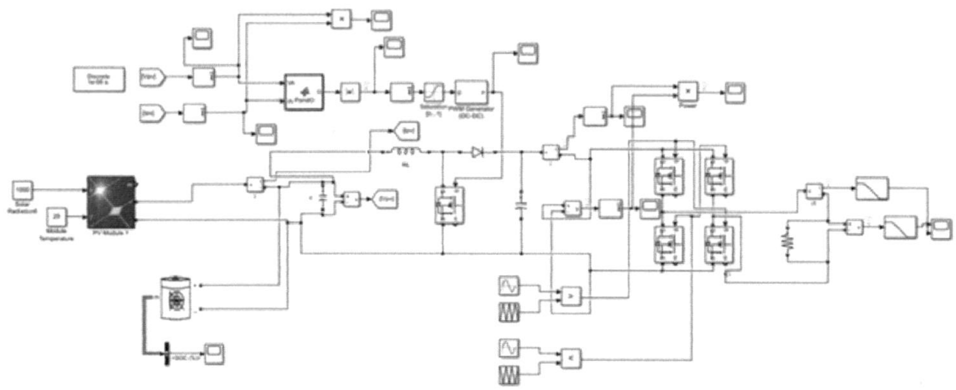

Figure 3. Simulink block diagram of PV power system.

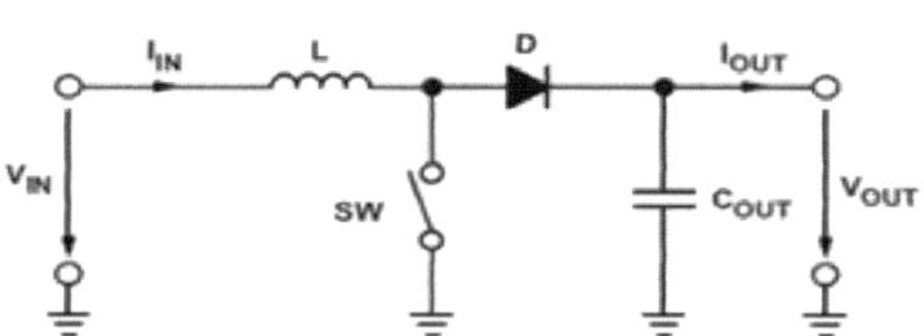

Figure 4. Boost converters.

Table 1. Design Specifications of Boost converter.

S. No.	Component	Specification
1.	PV Module	12Vp

doi:10.1088/1757-899X/1295/1/012012

2.	Fs	5kHz
3.	Vin	12V
4.	Vo	24V
5.	L_1	3.6mH
6.	C_0	0.345μF

Design of a boost converter specifications are shown in table 1. And the value of inductor and capacitor is calculated using equation (1) and (2) and the ripple current is considered as 20% [10].

$$C_0 = \frac{I_0 \times d}{\Delta V_0 \times f_s} \tag{1}$$

$$L = \frac{V_{in} \times (V_0 - V_{in})}{\Delta I \times f_s \times V_0} \tag{2}$$

where, I_0 is output current, d is duty cycle, f_s is switching frequency, ΔI is ripple current, V_0 is output voltage, V^{in} is input voltage.

The simulation is carried out for standard test conditions like 250°C and 1000W/m^2 radiation and the inverter output voltage, current and its THD is shown if figures 5 and 6 respectively. It is observed that the THD of inverter is around 7% for the tested result

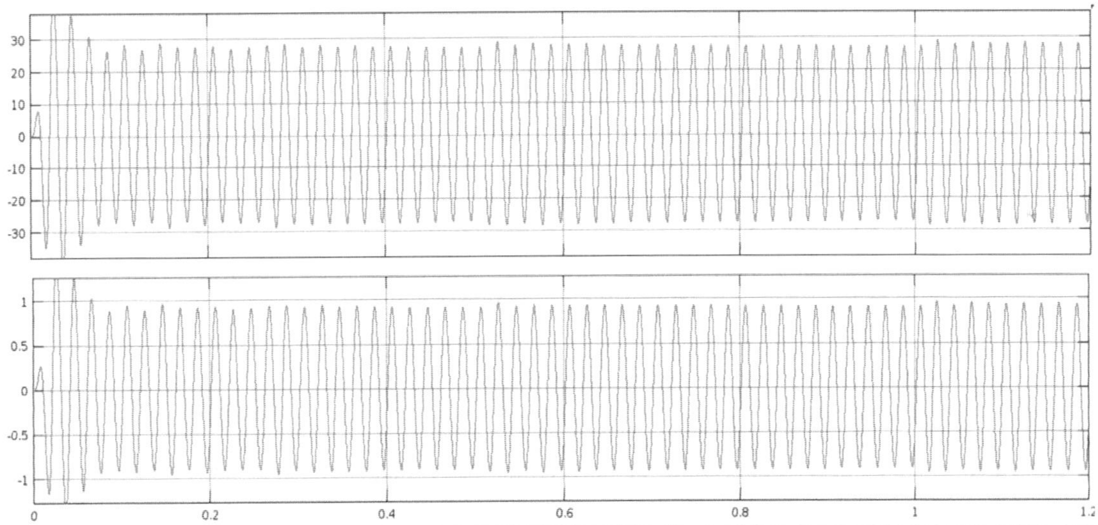

Figure 5. Inverter output voltage and current.

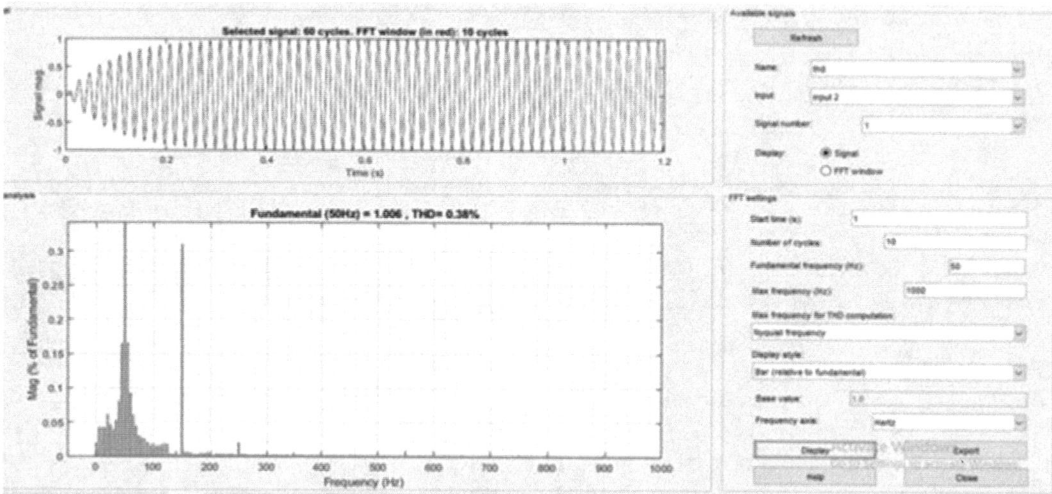

Figure 6. THD of inverter output current.

Figure 6 shows how the efficiency of the proposed and traditional systems varies throughout a range of output powers. Both methods were found to be relatively effective. As a result of the proposed system's additional boost converter losses, its overall efficiency is slightly lower than that of the standard system. The maximum efficiency is close to 50% of the nominal output power.

5. Hardware model

In this section the hardware & Software implementation aspects of the IoT-based Smart Power Monitoring System are explained. The hardware components, including Arduino Uno, voltage sensors, current sensors, and the PV module, play a crucial role in collecting data on power consumption and solar energy generation. The proposed system hardware is mostly broken down into three sections: the power circuit, driver circuit, and micro controller. With the aid of a micro controller adapter, the voltage-regulated supply for the micro controller is provided straight from the AC supply. The step-down transformer supplies the rectifier with power for the Driver circuit. The rectifier, output is connected to 12V regulator and then to the Driver ICs. With the aid of this 12V regulator, we can supply each driver IC with the 12V supply that it needs. Around the Driver circuit. It has a buffer IC. Between the driver IC and the microcontroller, this buffer IC functions as an intermediary circuit. The micro controller component with the Driver circuit is isolated by the buffer IC. It is employed to deliver constant pulses. Using this buffer IC, ripples or disturbances can be prevented if they occur. It needs a 5V supply for this buffer IC, which can be derived from the Arduino Uno microcontroller. It receives pulses from the Arduino Uno microcontroller, and these pulses are applied to the buffer IC and then to the driver ICs. The power circuit section now consists of an inductor, a boost converter, an H Bridge inverter, and other components. It is using a 12 V lead acid battery in case PV is not sufficient and battery is available, to supply the inductor with power for the inverter. The H bridge inverter receives this boosted output. Now that the inverter has converted the DC power to AC power, the load will be powered by this AC power. Three voltage sensors and two current sensors are used in the system. Figure 7 shows the image prototype of smart IOT integrated PV power monitoring system. Table 2 shows the hardware component details.

Figure 7. Hardware prototype.

Table 2. List of components used.

S. No.	Components	Specifications
1.	PV Panel	100 Wp, 22 Vmp, 4.52 Imp
2.	Rectifier (DB107)	1A
3.	Capacitor	1000uF,25volts
4.	Regulator12 V	12V
5.	IRF250N (4 no)	200V,30A
6.	TLP250H (Optocoupler IC)	12V,1.5A
7.	IN4007 Diode	700V,1A
8.	Transformer	230/12 V
9.	CD4050BufferIC	3-18V,0.32mA
10.	LED lamp	230V,9W
11.	Bridge Rectifier–BR1010	50V,10A
12.	Voltage sensor (3 no)	Robodo
13.	Current sensor (2no)	ACS712
14.	Resistors	100ohms,
15.	Inductor	0.8mH
16.	Arduino Uno Microcontroller	
17.	ESP 32 Wi-Fi module	

For real-time power monitoring, current and voltage sensors, as well as an ESP32 WiFi module for wireless communication and data transfer is used. The data analysis, visualization is achieved with the help of ThingSpeak [11]. The sensors measure the voltage and current of the batteries, the voltage and current of the loads, and the voltage of the PV panel. Figure 8 shows the sample code screen short of the system.

IOP Conf. Series: Materials Science and Engineering 1295 (2023) 012012 doi:10.1088/1757-899X/1295/1/012012

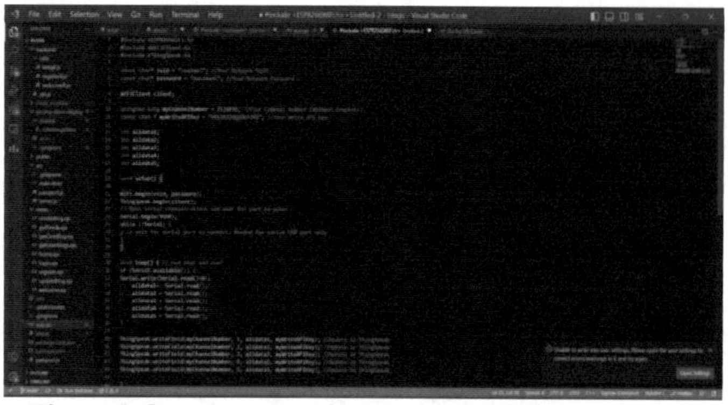

Figure 8. Sample code written to upload the data to cloud.

6. Results and discussion

The proposed system power monitoring at different stages is achieved using Thingspeak App and with the algorithm used to decide the effective utilization of PV or battery based on the decision. The prototype is tested for a sample lamp load. The results are graphically shown in Thingspeak App and the real time data and battery voltage is captured and is organized using a data wrapper. It displays charging and discharging patterns, battery voltage levels. Load voltage and current tables assess power delivery and stability, detecting load usage peaks or dips the solar voltage table reflects solar energy generation, analyzing this data help in system understanding, aiding, power management and decision-making. The sample data is shown in figure 9.

created_at	Load voltage	battery voltage	solar voltage	Load current	Battery Current
May 5, 2023 11:21:46 AM		48			
May 5, 2023 11:22:02 AM	46				
May 5, 2023 11:22:18 AM					53
May 5, 2023 11:22:37 AM					53
May 5, 2023 11:22:53 AM				48	
May 5, 2023 11:23:09 AM			53		
May 5, 2023 11:23:25 AM		48			
May 5, 2023 11:23:40 AM	46				
May 5, 2023 11:23:56 AM					53
May 5, 2023 11:24:14 AM				48	
May 5, 2023 11:24:29 AM			53		
May 5, 2023 11:24:56 AM				48	
May 5, 2023 11:25:12 AM			53		
May 5, 2023 11:25:27 AM		48			
May 5, 2023 11:25:43 AM	46				
May 5, 2023 11:25:59 AM					53
May 5, 2023 11:26:14 AM				48	
May 5, 2023 11:26:30 AM			53		
May 5, 2023 11:26:46 AM		48			
May 5, 2023 11:27:01 AM	48				

Get the data · Created with Datawrapper

Figure 9. The screen shot of the real time data captured from the prototype.

Figure 10 shows the solar output voltage during different period of the day. It is observed that during the 13:55 to 2:00 PM the highest voltage is obtained and also after 2:05 PM 2:10 one more peak voltage is obtained. During that time the output is directly by the PV module.

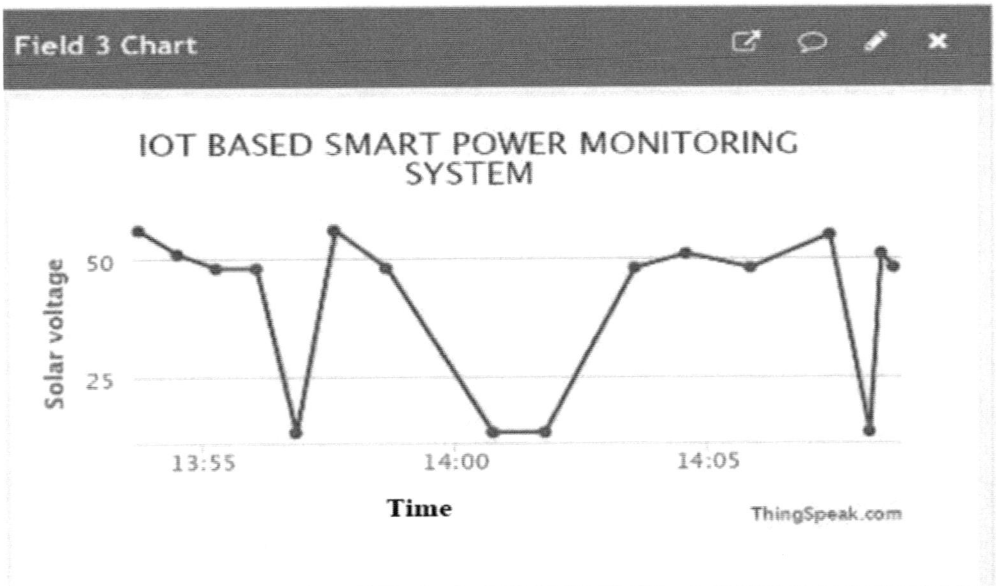

Figure 10. Dashboard display of real time PV voltage.

Figure 11 shows real time battery voltages. The graphical representation of these measurements helps to analyse and interpret the data effectively.

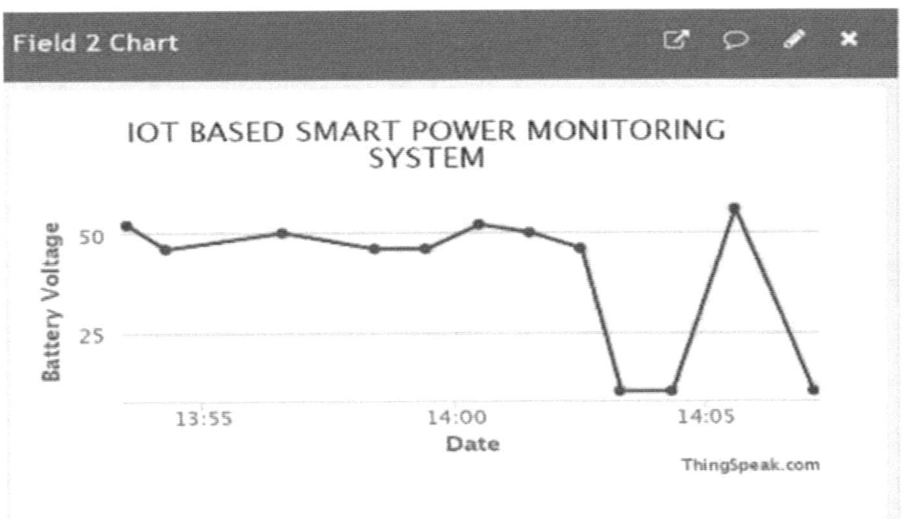

Figure 11. Dashboard display of real time battery voltage.

Figure 12 shows the battery current with respect to time a, the variation of battery current is observed between 13:55 PM to 2:05 PM. Depending on load the battery current is also varying.

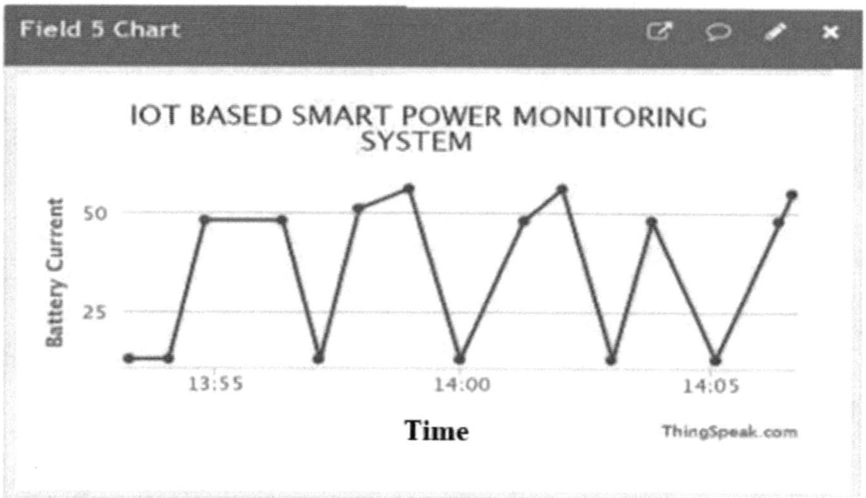

Figure 12. Dashboard display of real time battery current.

Figure 13 shows the variation of load voltage with respect to time. Solar or PV output voltage, battery voltage, battery current is all simultaneously obtained on the dashboard at the same time. It is clearly observed that when PV is there battery will not be discharged. Users can easily identify patterns, anomalies, and correlations within the voltage and current values, facilitating in-depth analysis and decision-making.

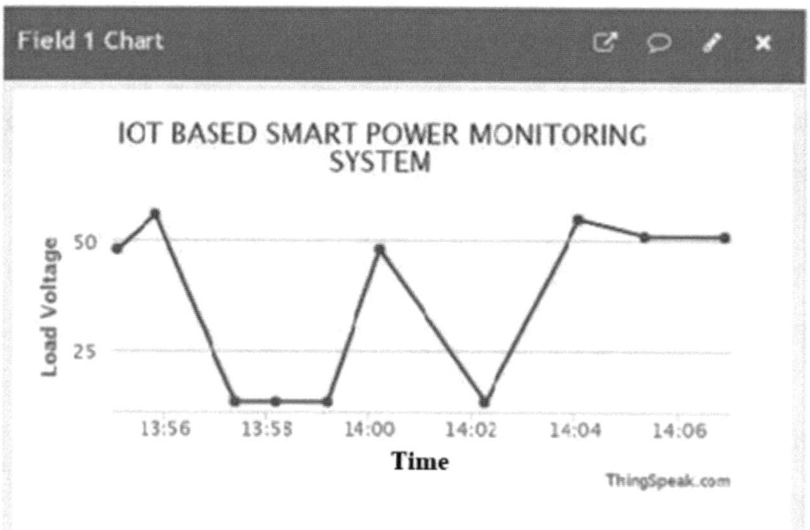

Figure 13. Dashboard display of real time load voltage.

7. Conclusion and future scope

In this paper an IoT-based PV integrated power monitoring system is designed, and hardware prototype is developed. It is a scale down model to understand the effective utilization of PV power and it's mentoring effectively with the help of IOT and ThingspeakApp. The integration of IoT technologies, such as the ESP8266 Wi-Fi module and cloud-based storage and analytics provided by ThingSpeak, enables seamless communication, remote monitoring, and control. Users can access the power monitoring system from anywhere and at any time, gaining valuable insights into their energy usage through intuitive data visualization and analysis. The paper objectives on power monitoring of data visualization, energy efficiency analysis, and integration with IoT technologies, energy conservation, and cost savings are all successfully achieved. The future scope includes advanced data analysis using machine learning for energy efficiency and expanded compatibility with devices through integration with smart home automation.

8. References

[1] Jafari M, Kavousi-Fard A, Chen T and Karimi M 2023 *IEEE Access* **11** 17471
[2] Tuyen N D, Quan N S, Linh V B, Van Tuyen V and Fujita G 2022 *IEEE Access* **10** 35846
[3] Bhau G V, Deshmukh R G, kumar T R, Chowdhury S, Sesharao Y and Abilmazhinov Y 2023 *Mater Today Proc* **80** 3697
[4] Xu S, Cao B, Chang L and Zhou J 2021 *IEEE J Emerg Sel Top Power Electron* **9** 5851
[5] Zafar S, Miraj G, Baloch R, Murtaza D and Arshad K 2018 *Engineering, Technology & Applied Science Research* **8** 4
[6] Wan Z, Song Y and Cao Z 2019 *Proceedings of 2019 IEEE 3rd Information Technology, Networking, Electronic and Automation Control Conference, ITNEC* p 18738137
[7] Gopal M, Chandra Prakash T, Venkata Ramakrishna N and Yadav B P 2020 *IOP Conference Series: Materials Science and Engineering* **981** 032037
[8] Pawar P, TarunKumar M and Vittal K. P 2020 *Measurement (Lond)* **152** 107187
[9] Rupesh M and Anbu Selvan N 2021 *Journal of Physics: Conference Series* **1964** 052001
[10] Khan F, Siddiqui M A B, Rehman A U, Khan J, Asad M T S A and Asad A 2020 *2020 International Conference on Engineering and Emerging Technologies, ICEET 2020* p 19493952
[11] Hasan M K, Ahmed M M, Pandey B, Gohel H, Islam S and Khalid I F 2021 *Wireless Communications and Mobile Computing* **2021** 6544649

Advanced Compensation Methods for Isolated Unsymmetrical Three-Phase Multiple Microgrids Employing Energy Storage System (ESS)

O Hemakesavulu[1], M Padma Lalitha[1] and C Haritha[1*]

[1] Department of Electrical and Electronics Engineering, Annamacharya Institute of Technology & Sciences, New Boyanapalli, Rajampet ,Annamayya, Andhra Pradesh, India

cheruvupalli.haritha4397@gmail.com

Abstract. The integration of unbalanced loads into three-phase microgrids (MGs) will be the emphasis of this study. The power quality difficulties that arise when the point of common coupling (PCC) is reached, will be the primary concern of this investigation. A few of these issues include unbalanced power flow, voltage, and the sharing of reactive power in an unbalanced manner. An energy storage system (ESS) that makes use of innovative control algorithms is used to correct the imbalances that occur in island multi-microgrids (MMGs). The reactive power generated by the MMG is balanced by the ESS's two types of control approaches, modified reverse droop control and multifunction control. The active and reactive power references needed by the ESS are likewise produced by these control methods. The proposed control algorithms ensure that PV inverters of varying ratings have optimal load power distribution, hence minimizing power imbalances and PCC voltage in the MMG. The MATLAB/SIMULINK application may be used to do an analysis on the outcomes of the simulation.

1. Introduction

Solitary-phase PV-MGs are employed more frequently in Multi-phase Multi-Module (MMG) systems, which leads to Unbalanced Voltage and Unbalanced Power Flows. These issues have a deleterious effect on power flows at common coupling terminals [1]. These problems prompted the creation of hierarchical event-based control, dispatchable optimum multi-objective control, reactive power compensation (RPC), and voltage control systems. Unbalanced MMGs can have their electricity balanced using single-phase PV inverters. Battery energy storage systems (BESSs) can be used with hybrid storage systems and techniques to reduce uneven voltage. Secondary control and droop per phase control are two control methods for unbalanced 3-phase, 4-wire MGs [2]. In droop control methods, leader-following, event-triggered voltage regulation, and fuzzy logic systems have all been used to low-voltage MGs connected to PV-DGs. Research on multiarea inverter-based MG power quality considers geographical dispersion, PCC, DG terminals, and sensitive loads. The author suggests using an electronic flow regulator to set up a hierarchical control system for power-sharing regulation across DGs and phases for secondary power in islanded S/T-MGs [3]. To reduce unbalanced power and improve reactive power sharing performance, this system employs a modified reverse droop control (MRDC) system made up of a power compensator, voltage compensator, and multifunction controller.

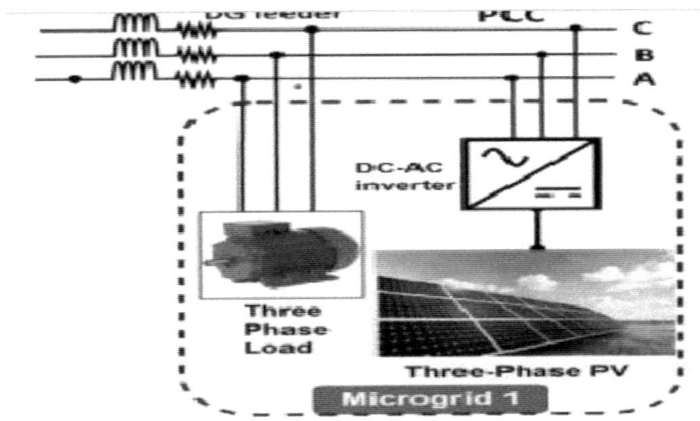

Figure 1. Conventional arrangement of the imbalanced MMG setup along with energy storage system (ESS).

This research discusses modern MRDC and multifunction controller for unbalanced S/T-MG power quality control methods for ESS (figure 1). These tactics are distinct and haven't received a thorough analysis in other sources. Unlike prior reverse droop controllers, the MRDC and multifunction controllers only evaluate power at the primary local controller. Reactive power sharing, three-phase PCC power balancing, and RPC are all possible uses for the multifunction controller. An investigation of the stability of the recommended controller inside the closed-loop control system for the imbalanced MMG is carried out under several different scenarios [4].

2. Existing method

The current architecture employs distributed control based on hierarchical events, an optimal multi-objective control that may be dispatched, and consensus control using a single-phase energy storage system (figure 2). Additionally, it uses virtual impedance control, droop control, a stochastic mixed-integer nonlinear programming approach, droop control per phase, and other secondary control methods [5].

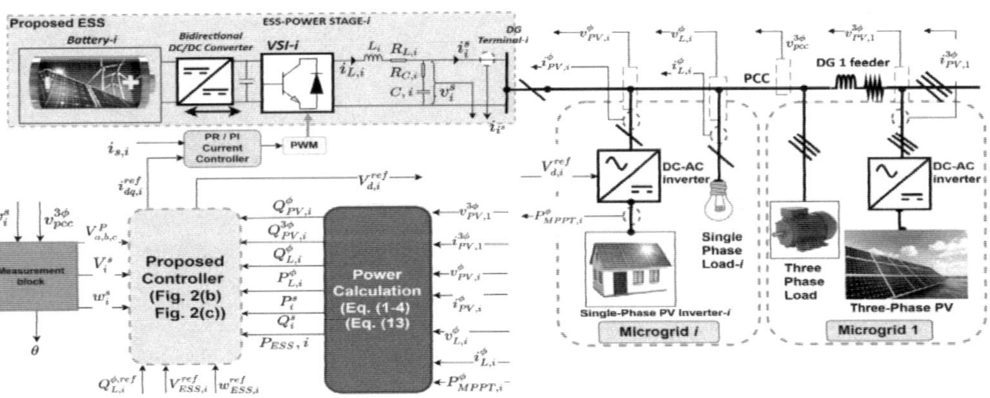

Figure 2. Block diagram of existing system.

3. Proposed method

The presented design uses contemporary technologies to manage an ESS. The three-phase PV employs a revolutionary reverse droop control technique to increase power efficiency and preserve the PCC voltage. In this method, a voltage compensator is produced by fusing a power compensator (PC) & a voltage compensator (VC). "In addition to reducing the uneven voltage of the three-phase PCC, the creative controller also handles uneven three-phase power and enhances reactive power distribution." This is made feasible by the creative controller's integration of a multifunctional controller (figure 3).

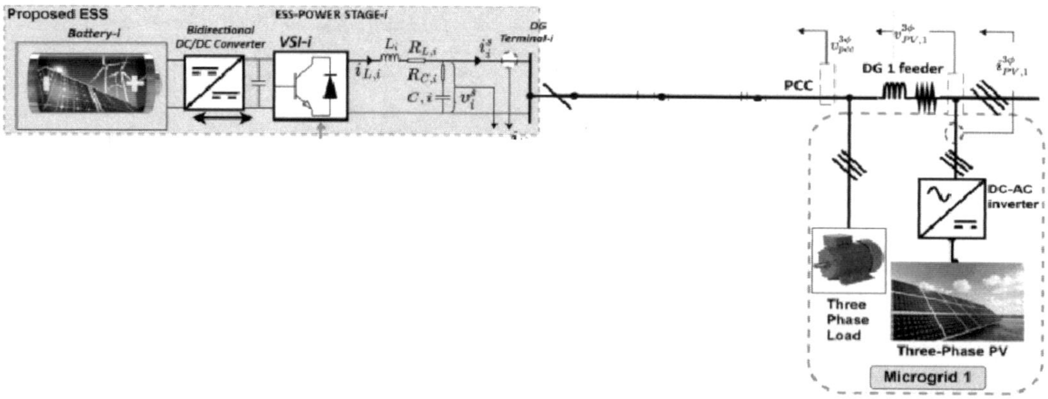

Figure 3. Block diagram of proposed system.

4. Simulation results

The schematic diagram of the proposed system is as shown in figure 3, an Energy storage system is connected to a Voltage Source Inverter (VSI) with the help of a bidirectional dc-dc converter connected to three phase load parallelly connected with dc-ac inverter connected to a PV system with the help of point of common coupling (PCC). Measurement block consists of phase locked loop, integrator etc., connected to multifunction controller scheme which consists of integrator, differentiator, PR current controller, VSI, current controller etc (figure 4).

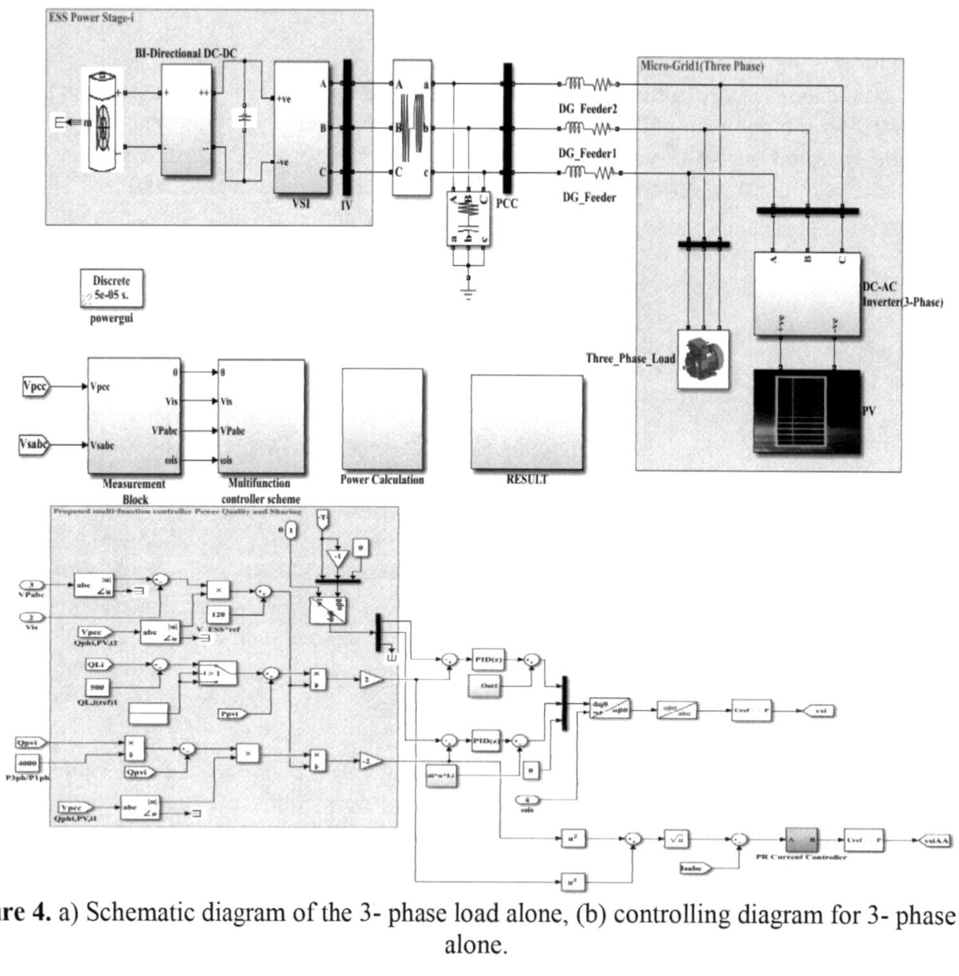

Figure 4. a) Schematic diagram of the 3- phase load alone, (b) controlling diagram for 3- phase load alone.

The following represents the output voltage of the point of common coupling (figure 5) before and after implementing the Reactive power sharing algorithm (RPSA) before and after implementation of the RPSA. The output voltage of the point of common coupling (PCC) is unbalanced because of the uneven power flow from the ESS to overcome this we use the power sharing algorithm after the implementation of this, the balanced power flows to the loads [6].

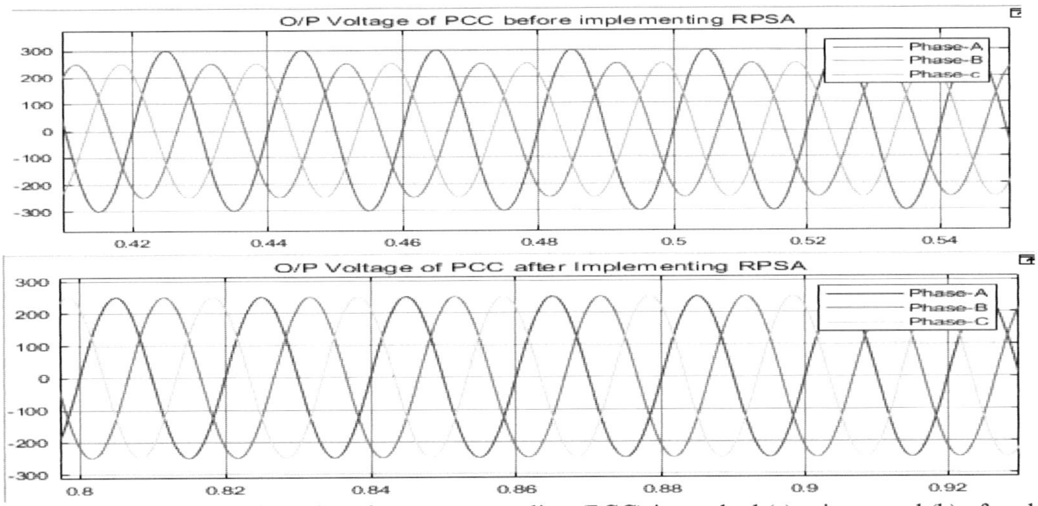

Figure 5. Voltage when the point of common coupling (PCC) is reached (a) prior to and (b) after the deployment of a rapid power quality assessment (RPCA) system.

The reactive power sharing performance of the 3-phase PV inverter when it is connected to the energy storage system is showing below figure 6. The reactive power of a three phase PV inverter is high when compared to the single-phase PV inverters reactive power. The reactive power of the 3-phase inverter is much more than thrice of the reactive power of a 1-phase inverter [7–11].

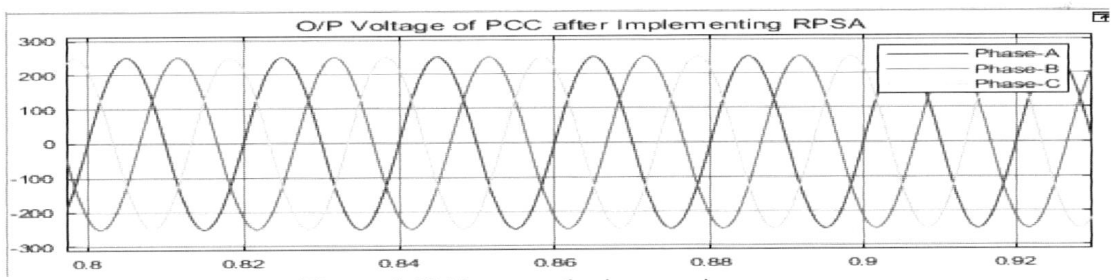

Figure 6. PV inverter sharing reactive power.

The active and reactive power of the 3-phase PCC terminals is shown as shown below figure 7. The power of the point of common coupling does not change with the load, there will be no variation in the active and reactive power because it is independent of phase of either the load or the PV inverters connected to the PCC [12–14].

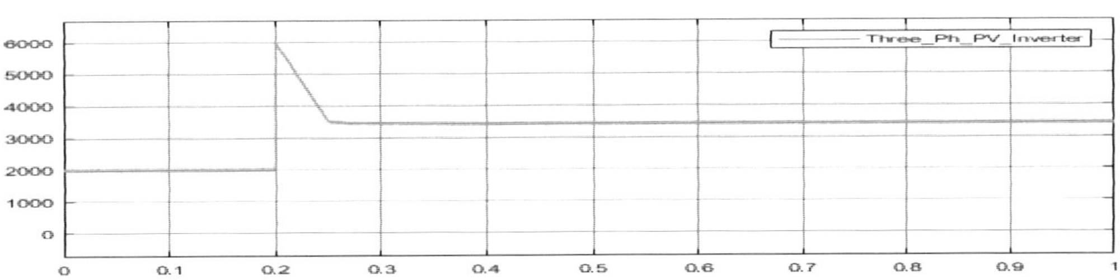

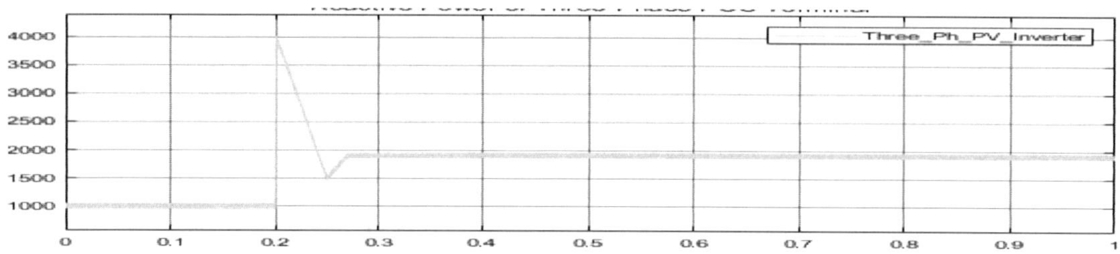

Figure 7. active power of 3-phase PCC terminal.

5. Conclusion

The presented design uses contemporary technologies to manage an ESS. The three-phase PV employs a revolutionary reverse droop control technique to increase power efficiency and preserve the PCC voltage. In this method, a voltage compensator is produced by fusing a power compensator (PC) and a voltage compensator. "In addition to reducing the uneven voltage of the three-phase PCC, the creative controller also handles uneven three-phase power and enhances reactive power distribution." This is made feasible by the creative controller's integration of a multifunctional controller.

6. References

[1] Vijay A S, Doolla S and Chandorkar M C 2020 Unbalance mitigation strategies in microgrids *IET Power Electronics* **13** 1687

[2] Kumar S, Yashaswini H K, Sharma N K, Bajaj M, Naithani D and Maindola M 2022 Classical Secondary Control Techniques in Microgrid Systems - A Review *Proceedings - 2022 2nd International Conference on Innovative Sustainable Computational Technologies, CISCT* p 22682768

[3] Akdogan M E and Ahmed S 2020 Control Hardware-in-the-loop for Voltage Controlled Inverters with Unbalanced and Non-linear Loads in Stand-alone Photovoltaic (PV) Islanded Microgrids *ECCE 2020 - IEEE Energy Conversion Congress and Exposition* p 20134931

[4] Zhou J, Xu Y, Sun H, Li Y and Chow M Y 2020 Distributed Power Management for Networked AC-DC Microgrids with Unbalanced Microgrids *IEEE Trans Industr Inform* **16** 1655

[5] Nejabatkhah F, Li Y W and Tian H 2019 Power quality control of smart hybrid AC/DC microgrids: An overview *IEEE Access* **7** 52295

[6] Nejabatkhah F and Wei Li Y 2019 Flexible unbalanced compensation of three-phase distribution system using single-phase distributed generation inverters *IEEE Trans Smart Grid* **10** 1845

[7] Wu D, Tang F, Vasquez J C and Guerrero J M 2014 Control and analysis of droop and reverse droop controllers for distributed generations *2014 IEEE 11th International Multi-Conference on Systems, Signals and Devices, SSD* p 14268670

[8] Vergara P P, Lopez J C, Rider M J and Da Silva L C P 2019 Optimal Operation of Unbalanced Three-Phase Islanded Droop-Based Microgrids *IEEE Trans Smart Grid* **10** 928

[9] Zhou J, Sun H, Xu Y, Han R, Yi Z, Wang L and Guerrero J M 2021 Distributed Power Sharing Control for Islanded Single-/Three-Phase Microgrids with Admissible Voltage and Energy Storage Constraints *IEEE Trans Smart Grid* **12** 2760

[10] Wang X, Qing H, Huang P and Zhang C 2020 Modeling and stability analysis of parallel inverters in island microgrid *Electronics (Switzerland)* **9** 463

[11] Akdogan M E and Ahmed S 2022 Novel Compensation Methods Using Energy Storage System (ESS) in Islanded Unbalanced Single-/Three-Phase Multimicrogrids *IEEE Transactions on Industry Applications* **58** 5457

[12] Zeraati M, Hamedani Golshan M E and Guerrero J M 2018 Distributed Control of Battery Energy Storage Systems for Voltage Regulation in Distribution Networks with High PV Penetration *IEEE Trans Smart Grid* **9** 3582

[13] Chua K H, Lim Y S, Taylor P, Morris S and Wong J 2012 Energy storage system for mitigating voltage unbalance on low-voltage networks with photovoltaic systems *IEEE Transactions on Power Delivery* **27** 1783
[14] Akdogan M E and Ahmed S 2021 Energy Storage System (ESS) for compensating unbalanced multi-microgrids using modified reverse droop control *Conference Proceedings - IEEE Applied Power Electronics Conference and Exposition – APEC* p 21173438

Microgrid Protection Challenges and Solution

Sangeeta Modi[1*], and P Usha[2]

[1] Department of Electrical and Electronics Engineering, PES University, Visvesvaraya Technological University, Belagavi, India
[2] Department of Electrical and Electronics Engineering, Dayananda Sagar College of Engineering, Bengaluru, Karnataka, India

smodi@pes.edu

Abstract. Microgrids are embryonic as inspiring solution to the various concerns such as environmental, economic, depletion of the resources for the fuel availability and power mismatch. Microgrids are going to become one of the core components of the upcoming power system. So, it is essential to understand various issues and challenges in microgrid. Power management, constant voltage and frequency, control of various distributed generators and Protection of the microgrid are major areas of concern. Not much work has been done on the protection side of the microgrid. Detecting fault and obtaining the appropriate setting of the protective gear for unconventional topology of the microgrid is very challenging. In this work a solution has been proposed and implemented in microgrid controller (MGC) for fault detection and classification in a microgrid. Effectiveness of proposed method is tested in MATLAB /SIMULINK environment.

1. Introduction

The latest expansion in the small-scale power generation with distributed energy resources and power electronic interfaces initiated the researchers to the concept of future power generation technologies such as microgrids. As per consortium for electric reliability technology (CERT) solutions of the USA the microgrid is defined as a micro power source and a low voltage system consisting of a cluster of loads, storage, and multiple distributed generators [1]. Microgrid operation is possible in two modes, and these are grid connected mode and islanded mode. The main advantage of microgrid emerges as a single controllable unit aiding it to deliver the cost benefits of large units [2,3]. Leony Ortiz et al presented simulation of hybrid microgrid in grid connected mode. Sohrab Mirsaeidi presented a comprehensive review on microgrid protection issues [4]. Protection of distributed network is no longer same as that of conventional grid network as there is a change in fault level because of change in topology [5].

Microgrids are also classified based on types of sources involved DC microgrid, AC microgrid and hybrid microgrid. DC microgrids, all the distributed energy resources and loads are connected to a common DC bus whereas in AC microgrids all the distributed energy resources are connected to a common AC bus. In hybrid microgrid various AC sources, DC sources and loads are connected to the respective AC and DC buses. The AC and DC buses are linked together with the help of power electronic interfaces.

The conventional grid has unidirectional flow of power from the transmission section to the loads whereas in microgrids power flow is bidirectional [5]. As the number of distributed generators or

renewable energy sources increases in the grid and both in transmission and distribution system, fault detection has become a major challenge in the microgrid. R. Sitharthan addressed protection issues in distributed generation based microgrid [6].
Figure 1 shows the structure of Hybrid microgrid. Solar and wind power source are connected to the DC bus with power electronic interfaces (PEI). Battery is connected to the DC bus with the help of PEI switch. Bi-directional interlinking converter is connected between the AC and the DC bus.

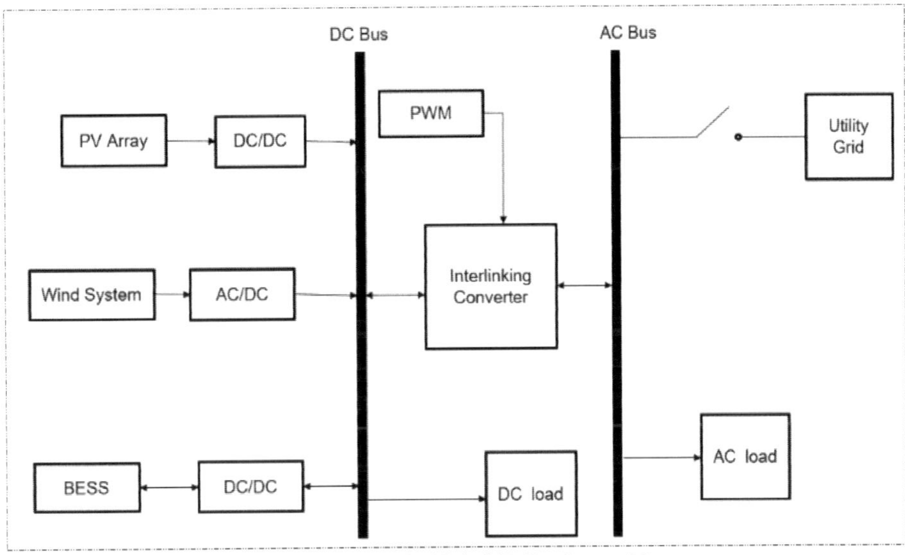

Figure 1. Hybrid AC/DC microgrid.

The protection scheme for Hybrid AC/DC microgrid system must be fast enough to detect the fault and protect the microgrid against various faults and associated consequences [7]. This is significant because the fault caused at inverter AC will cause temporary interruption with failure in communication and will cause stress on the converter [8]. Concept of smart overcurrent relay has been presented for the fault detection [9] whereas the need of intelligent AI based techniques in various control structures has been emphasised in [10].
Under fault condition, the MGC should be quick enough to detect the fault and then same data can be utilized for operating the associated protective gear in the system under consideration.
Some of the protection issues in microgrid systems are changes in dynamics of the microgrids based on the level of fault current, bi-directional fault current, false tripping, and blinding of protection, circulating current. Circulating current elimination technique has been discussed in reference [11]. For analysing any network, mathematical model is required which is discussed in reference [12]. Comparative analysis of pole to pole and pole to ground fault is presented and it is observed that pole to ground fault is difficult to detect as compared to PG fault [13]. Fault analysis in microgrid for developing an MGC is presented in [14]. Control strategies with associated dynamics and challenges of microgrid protection under various modes of operation are discussed in [15]. These challenges are discussed in detail and ML based solution is proposed and implemented in the later section of this paper. Classification results for various relays in a DC microgrid and necessity of adaptive feature of the protection scheme has been discussed by Jing Wang [16]. ML methods can handle extremely complex system and application of ML for power system network reliability, stability and security has been discussed [17]. In this work ML technique is applied for power system.
There are several protections schemes available for conventional power system but there is a need to devise a suitable protection scheme for the microgrids due to unconventional technology of the microgrid. In this work the application of machine learning algorithm is proposed and implemented for the microgrid controller (MGC) for fault detection which is suitable for both the modes of operation.

IOP Conf. Series: Materials Science and Engineering 1295 (2023) 012014 doi:10.1088/1757-899X/1295/1/012014

2. Microgrid challenges

There are various protection challenges and approaches to protect the conventional system. The conventional grid system allows unidirectional flow of power to the customers i.e., they operate radially. In such systems, the magnitude of short-circuit current is directly proportional to the fault current in the system. Therefore, protection scheme was provided with overcurrent-based protection devices.

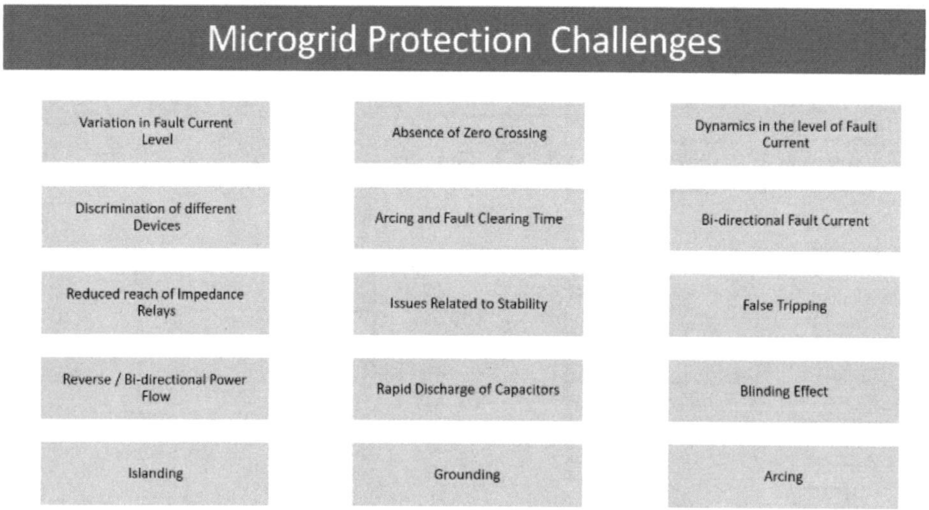

Figure 2. Microgrid protection challenges.

The conventional overcurrent-based protection devices can no longer be used for microgrids due to the various challenges posed by the microgrid such as Intermittence in generation of power, bidirectional flow of power, change in fault level etc. Figure 2 shows various protection challenges in microgrid.
The contribution of fault current to the inverter based distributed generators in MGs is only 2 to 3 times the maximum load current as the thermal competences of power electronic interfaces used is low. Therefore, the traditional overcurrent-based protection scheme can no longer be used as it may not trigger when the fault occurs in any section of the microgrid because of above mentioned challenges.
There are two types of faults in a DC system. One is Pole to Ground (PG) and the other one is Pole to Pole (PP). PP fault is a low impedance fault and causes severe damage to the system, but PG fault occurs frequently and affects the grounding system. The main objective of protecting scheme for grounding is the detection of fault and minimize the corrosion phenomena. Detection of solidly grounded system is easy as it has low ground impedance but the issue with the level of DC stray current. For ungrounded system, detection of fault current is challenging as they have low fault current. Protection gear operation involves arc phenomena. As DC currents lack zero-crossing current, interruption of current in DC section of the microgrid poses an issue in devising a protection scheme for AC/DC Microgrid.
Dynamics in fault current level - The amount of fault current that flows in the system varies as per the mode of operation of the microgrid. Additionally, the level of fault current changes as renewable energy based DGs are intermittent in nature. The magnitude of fault current when it is connected in grid connected mode and in islanded mode and is presented in the following figures 3 and 4. Fault current at fault point under grid connected mode when point of common coupling (PCC) is closed is given by equation (1) and (2).

$$I_f = I_{gf} + I_{g1f} + I_{g2f} \text{ (Grid Connected Mode - PCC Closed)} \quad (1)$$

$$I_f = 0 + I_{g1f} + I_{g2f} \text{ (Ilanded Mode - PCC Open)} \quad (2)$$

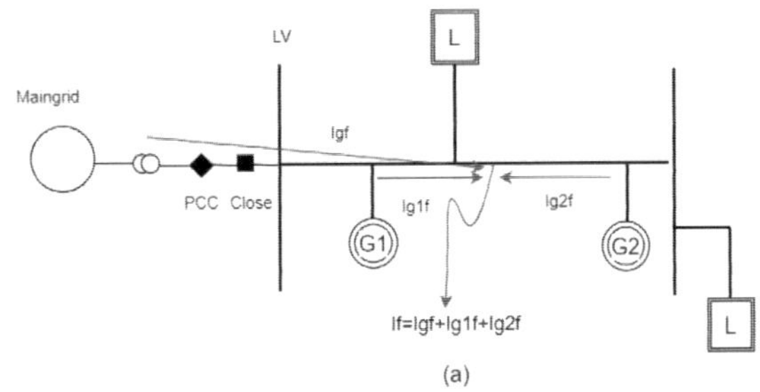

Figure 3. Fault current contribution-PCC closed.

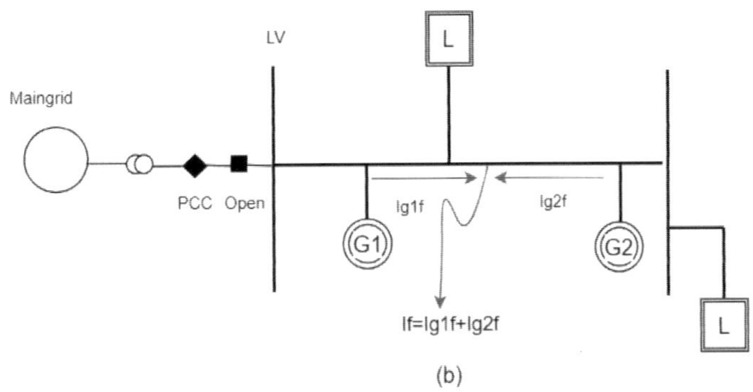

Figure 4. Fault current contribution-PCC open.

In figure 5, due to bi-directional power flow in a particular section of the system depending on location of the fault in the system, the level of the fault current changes and the direction of fault current alters. Hence, the traditional overcurrent relay can no longer be used for the protection of the microgrid. In figure 5, relay for circuit breaker 2 (CB2) will experience change in fault current direction when fault is set at F1 and F2. Relay setting is very challenging in this case.

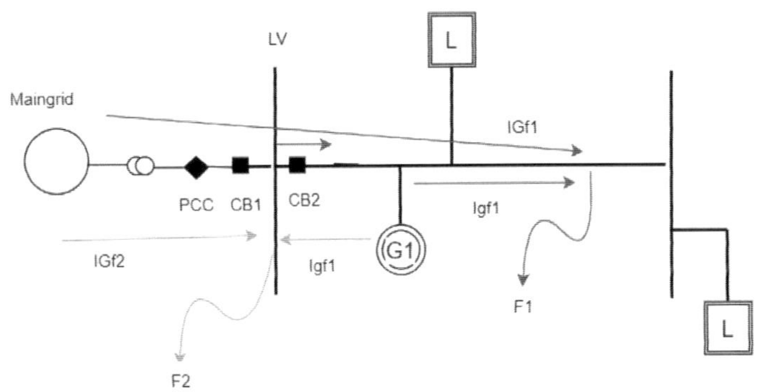

Figure 5. Bidirectional fault current.

Figure 6 below shows false tripping scenario of relay in different topology of the microgrid system. Feeder 1 will contribute to the fault current when there is a fault in Feeder 2 as shown below. If the fault current contribution of the DG exceeds the current setting, then CB2 is tripped before the operation of faulted feeder CB3. As a result, unnecessary power interruption will occur for loads connected to the healthy feeder.

Fault current in the system configuration shown in figure 6 is given by equation (3)

$$I_f = I_{DG} + I_{grid} \quad (3)$$

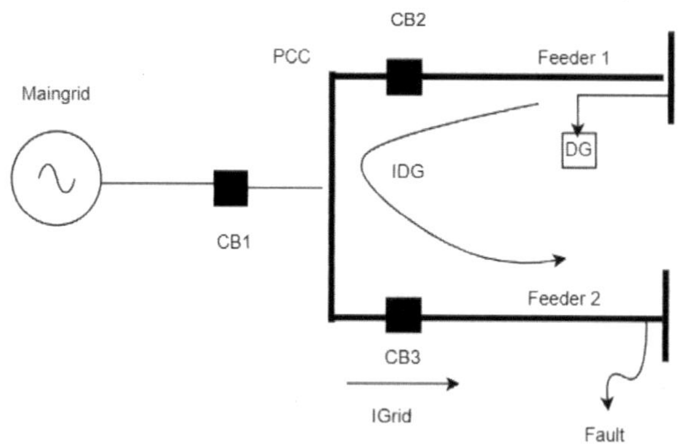

Figure 6. False-tripping.

Figure 7 shows another scenario of the system generally referred as blinding of protection. The contribution of the fault current is reduced due to the contribution of the DGs. As a result of this, the feeder relay is unable to detect the fault condition and this leads to blinding of protection. For all these above-mentioned reasons conventional protection techniques cannot be applied in hybrid microgrids. An intelligent protection scheme is required for fault detection in hybrid microgrid.

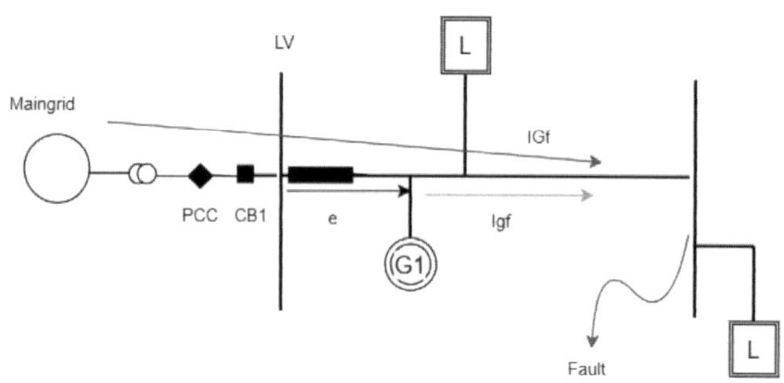

Figure 7. Blinding of protection.

3. Proposed solution and implementation

In this work an intelligent protection algorithm based on machine learning is proposed and implemented in microgrid controller (MGC) for detection and classification of faults in the hybrid microgrid. This algorithm is suitable for both the modes of operation of microgrid. Machine learning is the subclass of artificial Intelligence which permits the machine to learn exclusively. Supervised Learning is defined as a process where the labelled datasets are used to train the algorithm to either classify or predict the continuous value target function. Supervised learning is further classified as Classification and Regression based on the type of data mining.

Classification uses the labelled dataset to categorise into various datasets. The classification model has labelled data which comprises of both input and output. 'x' is the input variable and 'y' is the output variable shown in equation (4).

$$y = f(x) \tag{4}$$

The target feature is the function of the input variable x. The target feature 'y' is classification it is discrete and 'y' is regression when it is continuous. The classification model can be trained using various algorithms. 'x' includes input parameters i.e., voltage and current at various locations and 'y' is the predicted output which shows which type of fault has occurred in the system.

Unsupervised Learning is defined as an algorithm that learns from trends in the dataset without a supervision and predicts the unseen values. This algorithm is given an unlabelled dataset with input variable x without any out variable and it is supposed to analyse and find clusters inside. The unsupervised learning is further classified into clustering and association shown in equation (5).

$$y = f(p) \tag{5}$$

Semi-Supervised Learning technique (equation (6)) uses both supervised and unsupervised techniques i.e it learns from both labelled and unlabelled data.

$$y = (f(x), f(p)) \tag{6}$$

4. Methodology

In this work main objective is to detect and classify fault, Fine tree algorithm is implemented for MGC to detect and classify the fault. For data collection and training the following steps were followed as shown below.

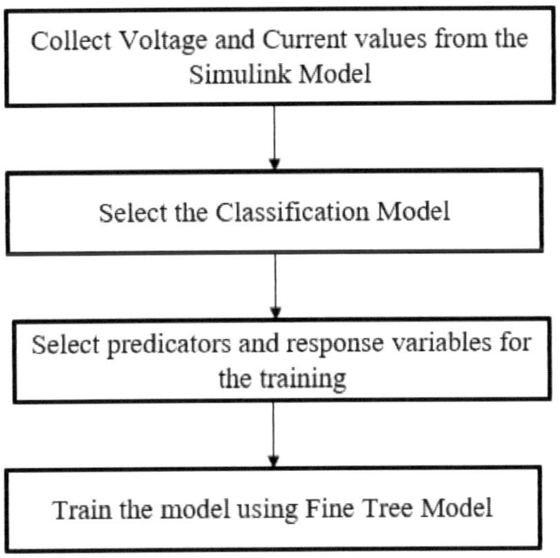

Figure 8. Basic steps in implementation.

Attribute Selection Measure-There are two ways of measuring attributes in decision tree and they are information gain and Gini index. Information gain is the measure of change in entropy after the segmentation of the dataset. The decision tree model always tries to maximise the value of information gain. Gini Index represents the represents the measure of purity or impurity while creating decision tree. Following equation is the mathematical representation of Gini Index shown in equation (7) [8].

$$Gini\ Index = 1 \times -\sum_{i=1}^{n} (pi)^2 \tag{7}$$

Confusion Matrix - Confusion matrix measures the performance of the model for the classified data in machine learning. Figure 9 shows the general representation of confusion matrix. The matrix has 4 different combinations of actual and predicted value. This is useful in calculating the fault detection and classification accuracy of the model shown in equation (8) [10].

$$Accuracy = \frac{TP + TN}{P + N} \tag{8}$$

where P= TP+FN and N= TN+FP; True positive (TP) and True negative (TN), False positive (FP) and False negative (FN)

IOP Conf. Series: Materials Science and Engineering 1295 (2023) 012014 doi:10.1088/1757-899X/1295/1/012014

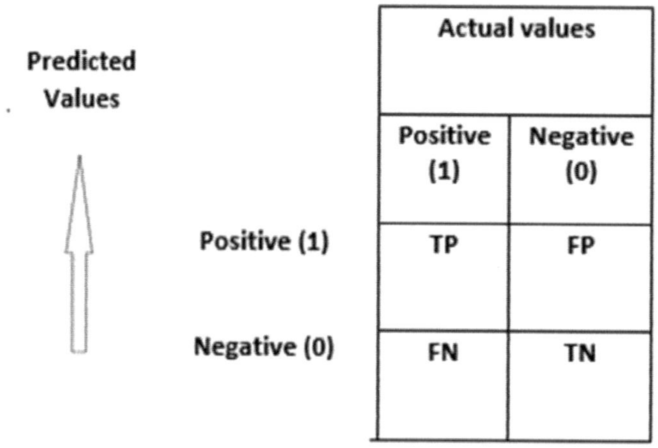

Figure 9. Confusion matrix.

5. Simulation results and analysis

The hybrid microgrid under consideration system is modeled and implemented in MATLAB/Simulink environment. The hybrid microgrid system has both AC and DC buses. The DC loads are connected to the DC bus and the AC loads are connected to the AC bus. The simulations were run for one second with sampling time 1e-06. Dataset was collected for the implemented hybrid microgrid system and model was trained using ML based Fine Tree algorithm. Different combinations of datasets are prepared, and their accuracy was measured and presented in this work. The trained model accuracy when PG and PP faults were created at PV after boost was 100%.

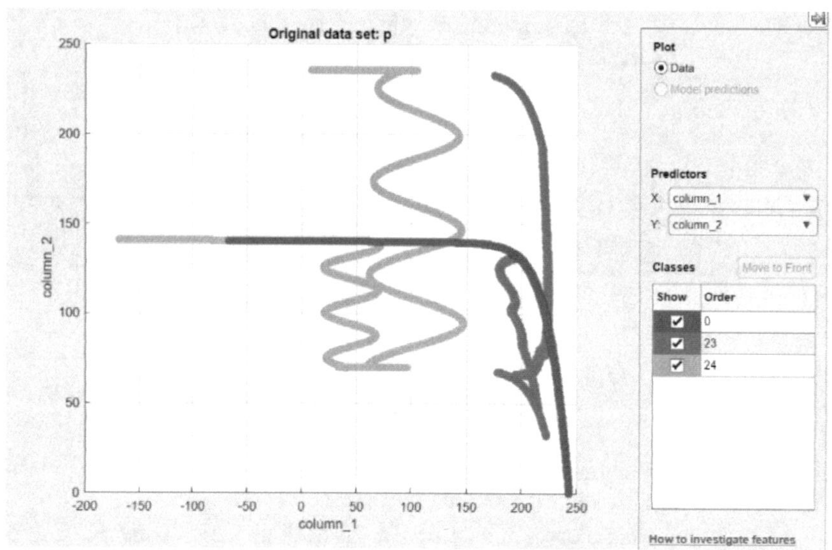

Figure 10. Scatter Plot for PG and PP fault created after PV boost converter.

The scatter plot in figure 10 shows the relationship between two predicators i.e., voltage and current. This helps in investigating features to include or exclude. The number 23 signifies PG fault at PV after boost, 24 signifies PP fault at PV after boost and 0 signifies the normal operation of the system. The confusion matrix in figure 11 evaluates the performance of the classification model. 691890 samples can identify the normal operation of the system, 672862 samples are able to identify PG fault at PV and 558574 samples are able to identify PP fault at PV. Similar observations were made when PG and PP fault at wind were merged. Figure 12 and figure 13 shows the scatter plot and confusion matrix for the wind system.

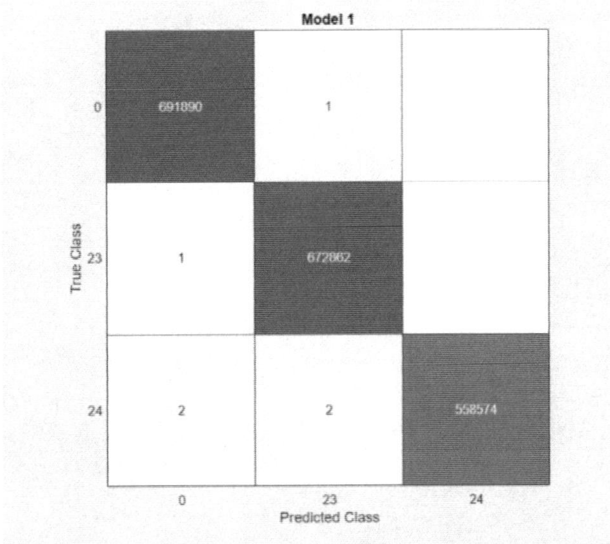

Figure 11. Confusion matrix for PG and PP fault at PV created after PV.

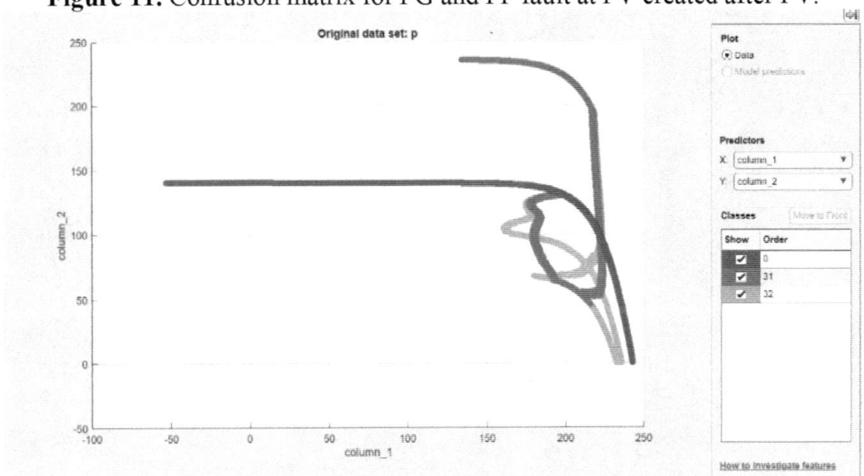

Figure 12. Scatter plot for PG and PP fault at wind before boost converter.

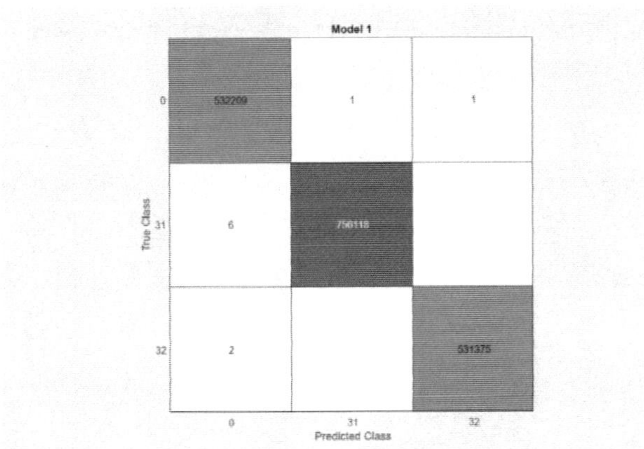

Figure 13. Scatter plot for PG and PP fault at wind before boost converter.

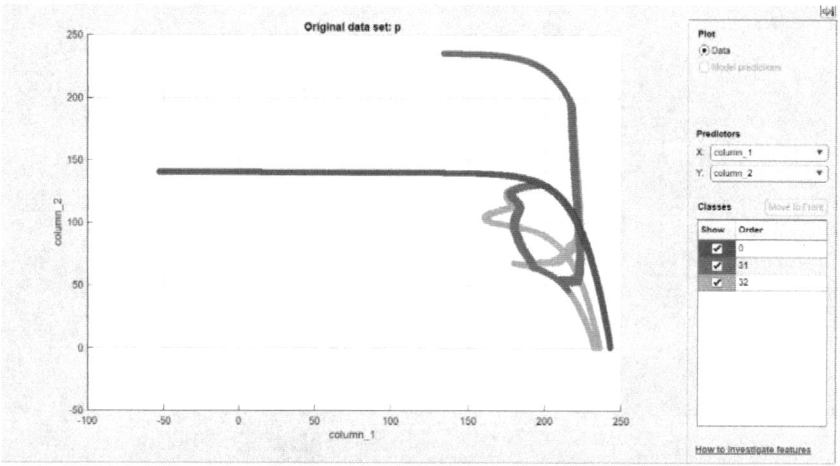

Figure 14. Scatter plot for PG and PP fault at wind before boost converter.

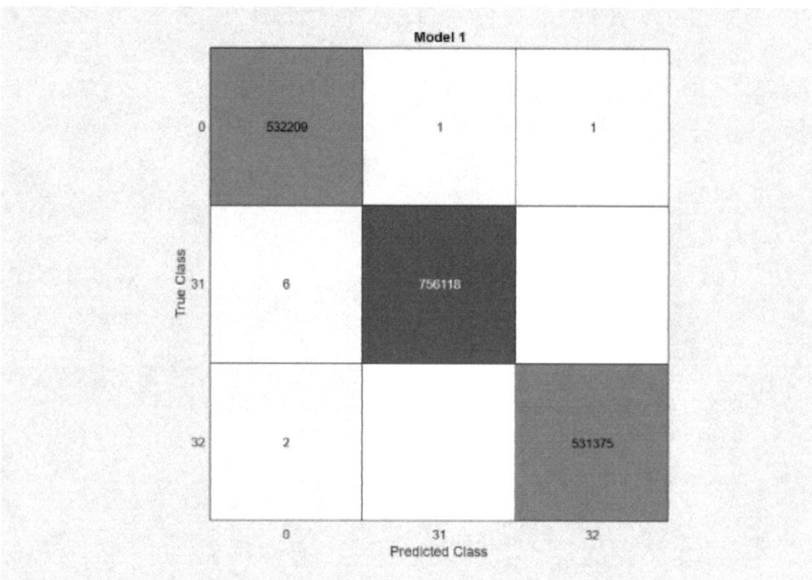

Figure 15. Confusion matrix for PG and PP fault at wind before boost converter.

When PG and PP fault at PV before and after boost were combined, the accuracy of the system was 100% shown in figure 14 and figure 15.

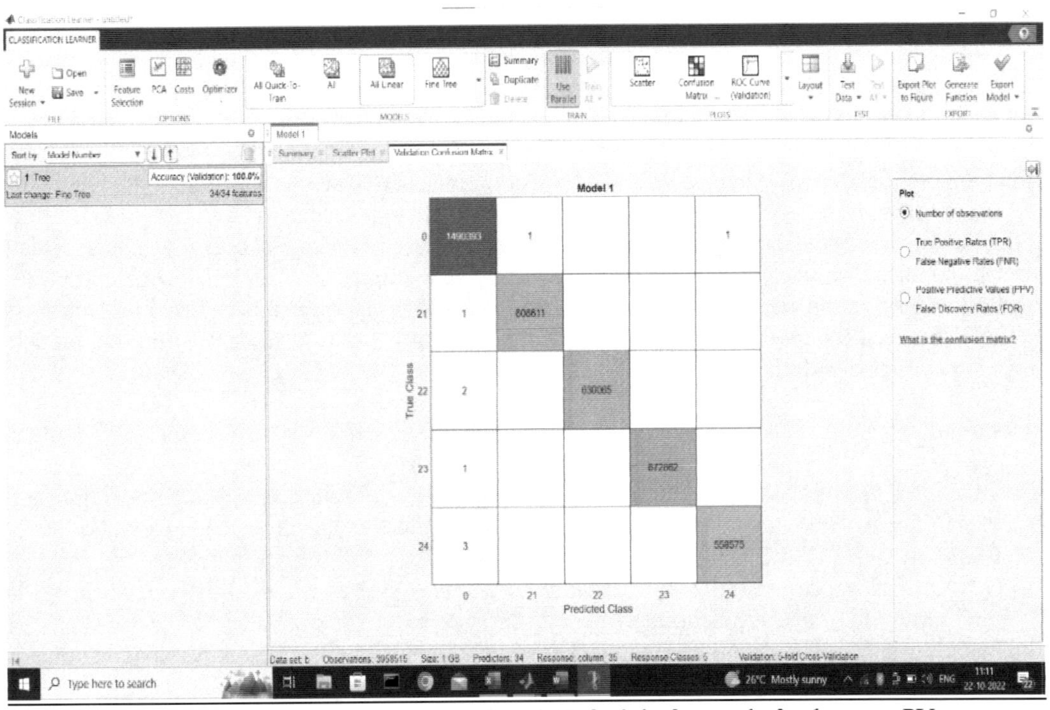

Figure 16. Confusion matrix for PG and PP fault before and after boost at PV.

Figure 16 shows the confusion matrix for the PV system. The number 0 signifies the normal operation of the system, 11 signifies PG fault at battery, 12 signifies PP fault at battery, 21 signifies PG fault before boost at PV, 22 signifies PP fault before boost at PV, 23 signifies PG fault after boost at PV, 24 signifies PP fault after boost at PV.

On similar lines, detection of faults at DC side of the hybrid AC-DC microgrid including all the faults at three different sources i.e., PV, wind, and battery. The accuracy of the trained model was found to be 93.1%.AC faults like LLLG fault, LLL fault and LG faults were created at AC bus. When LLLG fault, LLL fault and LG fault at supply were combined, the accuracy of the model reduced to 89.4.%. From the confusion matrix shown in Fig 18 When all the DC faults and AC faults are combined, the accuracy of the trained model reduced to 89.2% LG fault at AC side is considered as the most commonly occurring fault at AC side. The accuracy of the model with when DC faults and LG faults are combined is 90%.

Table 1. Simulation results.

S.No.	Fault Scenarios	Model accuracy
1.	PG & PP fault at PV after boost	100%
2.	PG & PP fault at wind power source before boost converter	100%
3.	PG & PP fault at PV before and after boost converter	100%
4.	PG & PP fault at PV, wind and battery	93.1%
5.	LLLG, LLL and LG fault at AC supply	89.4%
6.	PP, PG faults and AC faults	89.2%
7.	PP, PG faults and LG faults	90%

6. Conclusion

Protection schemes employed for a conventional power system cannot be applied for hybrid microgrid due to various reasons such as change in fault level, topology variation, intermittent nature of renewable energy sources, blindness of protection etc. In this work microgrid protection challenges are addressed. The proposed ML based algorithm is implemented for MGC and effectiveness of the same is tested in MATLAB/SIMULINK environment. Fault detection and classification accuracy obtained from proposed implemented algorithm with various scenarios has been presented in table 1. Further, this fault detection and location obtained from ML based algorithm is employed for operating the associated breaker in the microgrid. Accuracy of the proposed machine learning based fault detection and location algorithm has been found out and presented in table 1. As proposed algorithm is based on data driven technology, future work can be carried out in selecting the appropriate data for training the model to enhance the accuracy of the algorithm.

7. References

[1] Olivares D E, Mehrizi-Sani A, Etemadi A H, Cañizares C A, Iravani R, Kazerani M, Hajimiragha A H, Gomis-Bellmunt O, Saeedifard M, Palma-Behnke R, Jiménez-Estévez G A and Hatziargyriou N D 2014 Trends in microgrid control *IEEE Trans Smart Grid* **5** 1905

[2] Ortiz L, Orizondo R, Águila A, González J W, López G J and Isaac I 2019 Hybrid AC/DC microgrid test system simulation: grid-connected mode *Heliyon* **5** e02862

[3] Mirsaeidi S, Dong X, Shi S and Wang B 2017 AC and DC microgrids: A review on protection issues and approaches *Journal of Electrical Engineering and Technology* **12** 2089

[4] Kumar S, Yashaswini H K, Sharma N K, Bajaj M, Naithani D and Maindola M 2022 Classical Secondary Control Techniques in Microgrid Systems - A Review *Proceedings - 2022 2nd International Conference on Innovative Sustainable Computational Technologies, CISCT* p 22682768

[5] Zheng K H and Xia M C 2011 Impacts of microgrid on protection of distribution networks and protection strategy of microgrid *APAP 2011 - Proceedings: 2011 International Conference on Advanced Power System Automation and Protection* p 12675332

[6] Sitharthan R, Geethanjali M and Karpaga Senthil Pandy T 2016 Adaptive protection scheme for smart microgrid with electronically coupled distributed generations *Alexandria Engineering Journal* **55** 2539

[7] Wu Z, Zhao S, Hu M and Dou X 2012 Branch directional variation protection of AC microgrid *Zhongguo Dianji Gongcheng Xuebao/Proceedings of the Chinese Society of Electrical Engineering* **32** 158

[8] Cuzner R M and Venkataramanan G 2008 The status of DC micro-grid protection *Conference Record - IAS Annual Meeting (IEEE Industry Applications Society)* p 10290770

[9] Baran M E and Mahajan N R 2007 Overcurrent protection on voltage-source-converter-based multiterminal DC distribution systems *IEEE Transactions on Power Delivery* **22** 406

[10] Trivedi R and Khadem S 2022 Implementation of artificial intelligence techniques in microgrid control environment: Current progress and future scopes *Energy and AI* **8** 100147

[11] Modi S, Anand A and Usha P 2016 Microgrid control: A comparative study on control strategies for controlling the circulating current *International Conference Communication, Control and Intelligent Systems, CCIS* p 15872342

[12] Hivziefendic J, Vuic L, Lale S and Saric M 2022 Application of the Voltage Control Technique and MPPT of Stand-alone PV System with Storage *Advances in Electrical and Computer Engineering* **22** 21

[13] Modi S 2022 Microgrid Protection and Fault Analysis *Int J Res Appl Sci Eng Technol* **10** 4501

[14] Li Z, Lu J, Zhu Y and Jiang W 2017 Ground-fault characteristic analysis of grid-connected photovoltaic stations with neutral grounding resistance *Energies (Basel)* **10** 1910

[15] Ishaq S, Khan I, Rahman S, Hussain T, Iqbal A and Elavarasan R M 2022 A review on recent developments in control and optimization of micro grids *Energy Reports* **8** 4085

[16] Lazzari R and Piegari L 2023 Short-Circuit Protection Schemes for LVDC Microgrids Based on the Combination of Hybrid Circuit Breakers and Mechanical Breakers *International Transactions on Electrical Energy Systems* **2023** 9403058
[17] Wang J, Pinson P, Chatzivasileiadis S, Panteli M, Strbac G and Terzija V 2023 On Machine Learning-Based Techniques for Future Sustainable and Resilient Energy Systems *IEEE Trans Sustain Energy* **14** 1230

Modified Chameleon Swarm Optimization Algorithm to Improve the Power System Stability

R Shivakumar[1], and S Sowranchana[1*]

[1] Department of Electrical and Electronics Engineering, Sona college of Technology, Salem, Tamil Nadu, India

sowranchanasudhakar@gmail.com

Abstract. Power system stability has been difficult due to the occurrence of low frequency oscillation in the modern power system. Oscillation is mainly due to sudden changes of frequency, load, voltage, active power and reactive power. In this paper we provide a damping solution to oscillate the low frequency by using modified chameleon optimization algorithm. Damping performance and stability analysis of system is done by proposed modified chameleon Swarm Algorithm compared with conventional chameleon Swarm algorithm and Genetic algorithm with different operating condition in terms of Real power, Reactive power and Load disturbances. The comparative results confirmed that the proposed controller exhibit higher damping ratio, and better damping of deviations in speed and power angle to improve the stability of the system.

1. Introduction

Operation of modern interconnected power system with a certain operating condition in constant speed is difficult task to engineers due to the presence of disturbances with low frequency oscillation 0.2Hz to 2Hz which caused by the changes in load, real power and reactive power makes the system unstable [1,2]. Because of the sudden changes in the load, the modes get excited during operating conditions leads to transient's state of operation, it also refers to the power system's non-linear behaviour due to huge disturbances and low frequency oscillations which occurred in the system. Low frequency oscillation must be damped as quick as possible otherwise it may lead to many instabilities problem and even loss of synchronism in alternator. To overcome the low frequency oscillation and enhance the stability system is equipped with various power system damping controllers (PSDC) [3,4]. Implementing PSDC for various operation in multimachine power system is difficult and it requires complete analysis of Damping conditions with its parameters.

In recent years various approaches are considered to overcome the problem in PSDC that includes optimal, variable structure, adaptive and intelligent control. Despite this technique alternative controllers for damping are implemented by various approaches like fuzzy logic neural networks and neuro-fuzzy [5,6]. These techniques have a long optimization process which takes long computational time, and design is complex. In recent time algorithms are inspired by the behaviour of plants, flower pollination algorithm, and animal behaviour like elephant herding technique, which mimics the nature sound with a strategy. Various nature inspired algorithm is implemented in the system to overcome the problems in system. These techniques consist of particle swarm algorithm, butterfly optimization algorithm, genetic algorithm (GA), grasshopper algorithm, whale optimization algorithm. A new improved swarm algorithm is proposed to deal with fitness problems effectively. Chameleon optimization algorithm is the new innovative technique which belongs to swarm-based algorithm [7,8]. Oscillations with undamped frequency are damped by using power system dynamic controller like GAPSDC, CSAPSDC, MCSAPSDC. The damping ratio and error deviation of the GAPSDC is compared with CSAPSDC, MCSAPSDC. The MCSAPSDC used in the interconnected power system, it can overcome the low oscillation frequency by damping it and improve the system stability. The comparative analysis of various algorithm is specified in this paper.

2. Problem statement

In this chapter, modelling of three machine nine bus multi machine system is done and the operation of PSS3B in dual-input is operated with power system model.

2.1. Modelling of multi machine power system

Structure of IEEE multi machine system is given in figure 1. which represents the modelling and system analysis. Group of many interconnected systems with different power limits are taken to form a multimachine model.

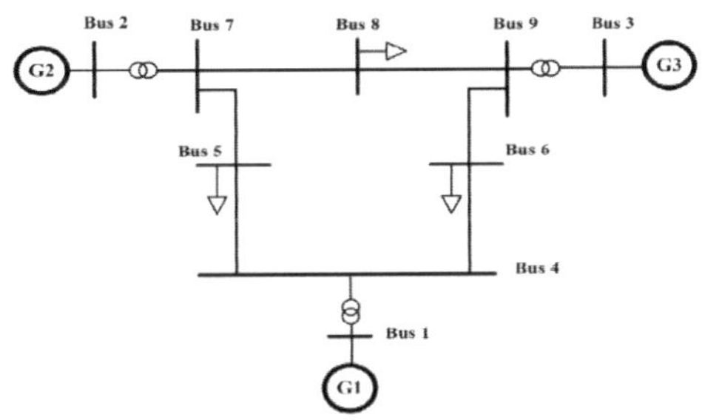

Figure 1. 3 Machine 9 Bus IEEE multi-machine model.

Modelling of synchronous generator is done by Heffron-Phillips generator model which are implemented for the analysis of power system [9,10]. PSDC executed in the system consist of both excitation system and generator state space model. system stabilizers with damping controllers provides the excitation system with desired damping conditions.

2.2. Damping controller structure

Figure 2. represents the power system damping controller structure (PSDC). Speed deviation (Δω) is the input of controller and the controller output (ΔUE) is provided as the feedback loop for the generator system.

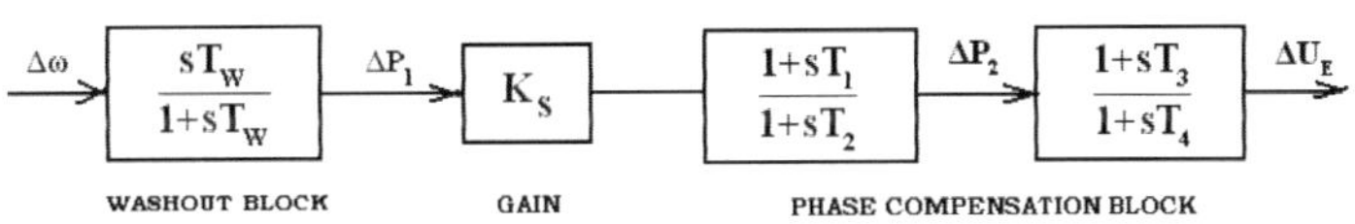

Figure 2. Structure of damping controller.

2.3. Dual-input power system stabilizer

PSS3B consists of two inputs to calculate the power angle and speed deviation of rotor and generator [11,12]. It has TS1 and TS2 with TW1 and TW2 are the time constant of wash out block and time constant of phase compensator.

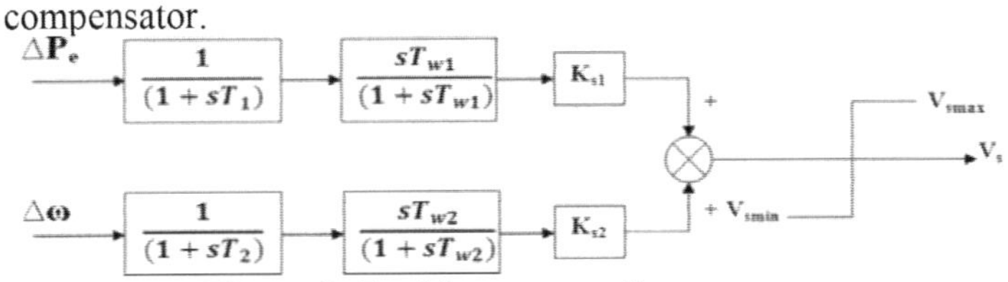

Figure 3. Dual input controller structure.

The system which includes speed of rotor angle deviation (Δω=ω_mech-ω_0), where electrical power is denoted by (ΔPe), accelerating Power (ΔPa), and frequency(Δf). Figure 3. represents the structure of dual input power system stabilizer. Equation (1) denotes the speed deviation of generator is proportional to the accelerating power (ΔP_m-ΔP_e). ΔP_m creates the oscillatory modes which can be implemented in the Power system damping controller [13,14] low pass filter is used to remove the low frequency in damping controller of the system.

2.4. Representation of power system- stabilizer

State space model of multi-machine system is represented in figure 4. The variables of open and closed loop systems are implemented in the power system modelling.

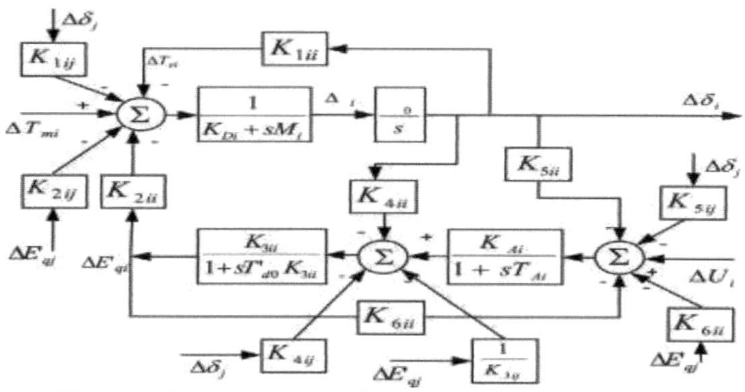

Figure 4. Representation of power system in state space.

3. Optimization criterion for power system

Stability of the power system are improved by solving the optimization problem with various limits to improve the system stability. Here, the limits which is based on the error deviation and damping ratio with certain limits are specified in equation (5) and (7). It is represented as objective function (P1) based on Damping ratio and objective function (P2) based on error deviation [13,14].

3.1. Objective function(P1)- based on damping ratio

Criterion of damping ratio is given by equation (1):

$$[P_1] = Min[\varepsilon_j], \varepsilon_j^e \varepsilon EMODE \quad (1)$$

Where, εi is the value of damping ratio, εEMODE damping ratios of all set of eigen values. It can be represented by the complex eigen values (σ± jω) and the damping ratio is given by equation (2).

$$[\varepsilon_j] = \frac{-\sigma}{\sigma^2} + \omega^2 \quad (2)$$

Weakly damped oscillation in a system will have a minimum value with a damping ratio [Min(εi)in a particular condition. And criterion here is to maximize[p1] the oscillations and to improve the system stability.

3.2. Objective function(P2)- based on error deviation

Error deviation of speed and rotor angle is denoted by e(t). P2 is the error deviation which is specified in the system, large error can be minimized when compared to the small errors with the integration and it is based on the integrate of square with error. To reduce the error deviation with integral square are main objective of P2. Oscillation should be settled quickly for the better performance of the power system.

4. Proposed and improved CSA algorithm stability

4.1. Overview of a conventional CSA

Braik has introduced a recent metaheuristic optimization technique Chameleon optimization in 2021. It is inspired by the food searching and hunting behaviour of the chameleon in respect to surrounding environment chameleon changes its colour with a unique manner. It can survive in semi-deserts, mountains, lowlands, deserts. The hunting behaviour of chameleon consist of different steps which includes prey tracking, chase by using eyes, and attacking it. The modelling of this techniques is given below.

4.1.1. Initialization and evaluation of function. Optimization process is performed by population-based CSA technique which can be generate the initial population randomly. The size of population initialized by n with a dimension search area m, and during the optimization problem with solutions are consider with all the individuals. In search area the location of chameleon with iterations are given in equation (3).

$$P_l^i = \left[P_{h-1}^i, P_{h-2}^i \cdots P_{h,m}^i\right] \tag{3}$$

i=1,2.... l represents the No of iterations, $P_{h,m}^{i}$ shows the chameleon position. The initialized population which depends on the number of chameleons and equation (4) denotes the dimension in the search space where it is denoted by equation (4).

$$P^i = b_l + r(b_u - b_l) \tag{4}$$

Initial vector can be denoted by pi of ith chameleon, r is a uniform random number in a range of 0. bu and bl are the lower and upper bound in search area. The solutions are calculated in every step which is based on the objective function.

4.1.2. Searching of Prey. Chameleon's behavior movement and updating position procedure is given in equation (5).

$$P_{h+1}^{i,j} = \left\{P_h^{i,j} + S_1\left(S_h^{i,j} - G_h^i\right)r_2 + S_2\left(G_h^i - P_h^{i,j}\right)r_1 \quad r_i \geq S_S\right\} \tag{5}$$

$$\left\{P_h^{i,j} + \mu\left(\left(b_u - b_1\right)r_3 + b_l^i\right)\operatorname{sgn}\left(rand - 0.5\right)r_1\right\} \quad r_i \leq S_S \tag{6}$$

Where in equation (5) h and (h+1) represents the iteration of hth and (h+1)th iteration count. ith chameleon with jth dimension dimensions is represented as i and j. Position of current chameleon is denoted as b_(h+1)^(i,j) . It has two positive numbers S1 and S2 where it controls the ability of exploitation. And the random numbers are represented as r1, r2, and r3 with ranges from 0 to 1. Ss is probability of chameleon for prey recognition. sgn(rand-0.5) represents the direction of exploration and exploitation with a value of 1 or -1. μ is an iteration parameter function which decreases according to the iteration numbers.

4.1.3. Eyes rotation behavior of chameleon. Location of the prey can be determined by chameleon with the ability of rotating their eyes in 360 degrees to spot and recognize the prey. It is formulated mathematically to find the hunting mechanism of an chameleons. The chameleon will change the location with respect to the prey location.

4.1.4. Hunting prey. Chameleon will strike the prey if it comes to the closer contact. The chameleon which is near to the prey is selected as the best chameleon and with individual optimum. By using its tongue, it attacks the target, and updates the optimum position, length can be doubled in the tongue to attack the prey. It also assists the best individual by expanding its search space to attack the prey effectively.

4.2. Proposed modified CSA for stability

The conventional CSA algorithm it fails to achieve the optimal values, and imbalance takes place in both exploration-exploitation phase which cause a weakness of strategy in exploring the search space. Modified chameleon swarm optimization is proposed to eliminate the shortcoming of conventional CSA by three different strategies and to avoid the complex real-world problems.

Improvements are considered in this system which includes the non-linear transfer operate to improve balance between exploitation and exploration. To avoid the convergence and stagnation it is provided with levy flight parameter for randomization. Algorithm is based on the performance of exploration and exploitation, imbalance between exploration and exploitation leads to a failure in global optimum. Where the first part of iteration will be exploration and second part are exploitation.

Levy fight, better in exploring search space and that parameter is Rand that can generate randomly in search space. Randomization of new parameter in levy flight can be obtained by combining both LF and Rand. Levy flight (RL) parameters are investigated in the search space. Algorithm of modified chameleon swarm algorithm is given in [6]. This particular algorithm is chosen because in compared with the several optimization techniques like linear and non-linear formed to solve the various optimization issues in science and technology. Searching and exploring are the major advantage in the MCSA algorithms. It also eliminates the outcoming of CSA by different methods and solve problems in real-world. These parameters are mainly used to optimize the damping controller parameter, to improve the stability of the system. Proposed parameters will help the chameleons sear in search space to obtain the balance between exploitation and exploration phase for more results.

4.3. Genetic algorithm-overview

In this section, the performance of CSA and MCSA damping controller are compared with the Genetic algorithm-based controllers. Genetic algorithm is a Bio inspired algorithm which has a genetics and natural selection. It consists of four essential operators that creates the individuals with fittest solution: selection, crossover, mutation, and replacement.

First essential operators are the selection of two parent chromosomes for reproduction in the initial population. Concept of roulette wheel is implemented in this paper. Selection operator will recombine and compared with other methods of selection.

Second essential operator is taking two parent chromosomes that selected in the selection process to produce a good offspring. Uniform crossover is considered in this paper. When crossover is completed, the mutation process will be takes place, and the lost genetic materials can be recovered in this process. It also involves the flipping of a bit that changes from 0.1 to 1.2 and vice versa.

The last stage of the genetic cycle is replacement. In this process the replacement of weak parent that involves with a generation gap of 0.6.

5. Simulation outcome- analysis of stability

In this work the damping ratio of close and open loop of a system can be calculated, and examined briefly. And the performances of various damping controllers like GAPSDC, CSAPSDC, MCSAPSDC are discussed and compared for the improvement of system stability.

5.1. Power system simulation- open loop

Simulation modelling, computations are done by MATLAB programming in different operating conditions. In this paper which is equipped with the generator 1(120 MVA, Voltage 13.6KV), generator 2 (190 MVA, Voltage 16.8KV) and generator 3 (160 MVA, Voltage 15.2 KV), generator 1 is taken as an infinite bus. Table 1 shows the damping ratio of open loop system. Operating disturbances of system are denoted by Pd. damping ratio in table are negative and shows the system are in unstable conditions.

Table 1. Damping ratio of weekly damped oscillation in open loop.

S.No.	Operating Condition (p.u)	Generator	Damping ratio
1	P=0.87,	G2	-0.0034532
	Q=0.03,		-0.0045027
	Pd=0.004	G3	
2	P=0.96,	G2	-0.0032376
	Q=0.08,	G3	-0.0041675
	Pd=0.009		

Response of speed deviation in open loop system is shown in figure 5. with the operating conditions shows the operating conditions P=0.87, Q=0.03, Pd=0. 004. Response of open loop system shows that the waveforms are continuously growing and oscillates for the long time. So, this condition of a system requires a damping controller to damp the oscillations in system with feedback.

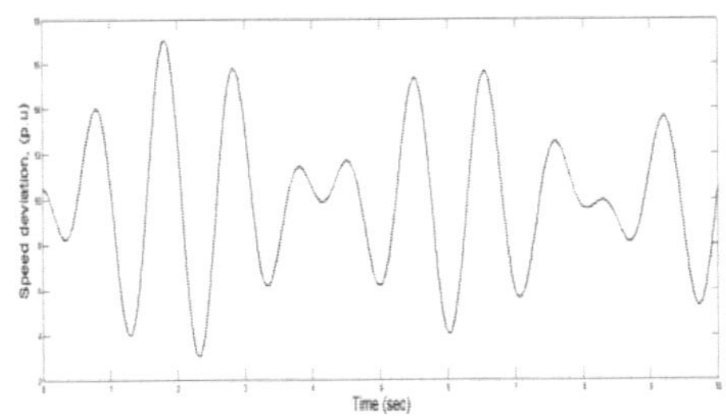

Figure 5. Speed deviation in open loop system without PSDC.

5.2. Proposed MCSA controller- execution

Simulation GAPSDC, CSAPSDC, and MCSAPSDC controller that have been installed in generator with the conditions to overcome the oscillations of low frequency. Various parameters of GAPSDC, MCSAPSDC are given in table 2.

Table 2. Variables used for GAPSDC, MCSAPSDC.

S.No.	GA variables		Damping ratio	
1.	Population Size	100	Population Size	100
2.	No of variables	03	No of variables	06
3.	No of iterations	100	No of iterations	100
4.	Cross over rate	10	Positive parameters	
			S_1	0.24
			S_2	1.45
			B_1	1.75

		B_2	1.75
5. Mutation rate	0.12	Φ (constant)	2.5
6. Generation gap	0.75	RL (random levy flight)	[0,1] 2.3

The control values and closed loop damping ratios of a system are shown in table 3. Damping ratio with minimum values is maximized by using objective function P_1, and makes the system stable. When compared to the three damping controllers, the proposed MCSAPSDC has a better performance in the power system. The response of time domain controllers which is same as the non-linear system with time. Response of power angle and speed deviation of system are shown in figure 6 and 7, with conditions Real Power, Reactive power and load disturbances (P=0.87, Q=0.03, Pd=0.002 p.u). where the waveform of MCSAPSDC will damp the overshoot in the system and solve it quickly by creating the deviation while compared with GAPSDC, CSAPSDC. It satisfies the objective function P_1.

Table 3. Damping ratios of weakly damped oscillation.

S.No.	Operating condition	Generator	MCSA- established with calculated optimal point Damping ratio of weakly damped oscillation Control parameters (damping threshold[ε_r]=0.04)				
			K_{sa1}, K_{sa2}, T_{s1}, T_{s2}	R, L	GAPSDC	CSAPSDC	MCSAPSDC
1.	Condition 1:		4.4620				
	P=0.87, Q=0.03,	G_2	4.0872	(0.5,1)	0.01967	0.012234	0.03967
	Pd=0.	G_3	0.8654	0.26	0.03456	0.034568	0.03654
	004p.u		0.7864				
2.	Condition 2		37.6547				
	P=0.96, Q=0.08,	G_2	17.4928	(0,1.5)	0.0543654	0.0786433	0.125681
	Pd= 0. 009	G_3	0.4354	0.27	0.0764873	0.0643269	0.167543
	p.u		0.1123				

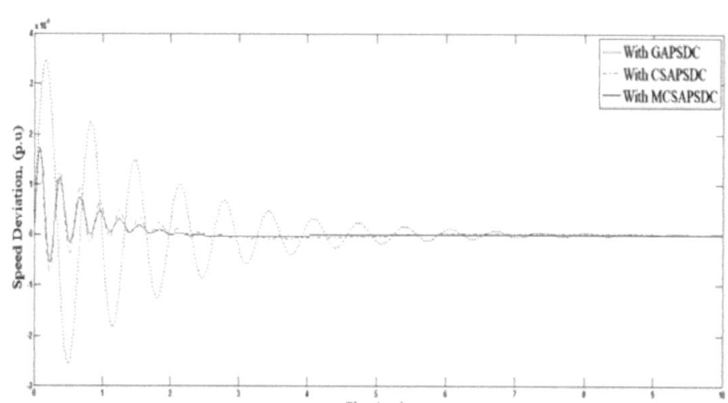

Figure 6. Speed deviation in G2 for the condition P=0.87, Q=0.03, Pd=0.002 pu.

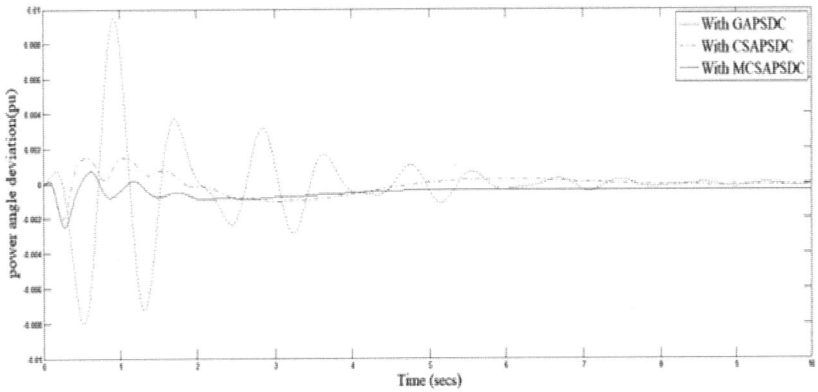

Figure 7. Power angle deviation in G_2 for the condition P=0.87, Q=0.03, Pd=0.002 p.u.

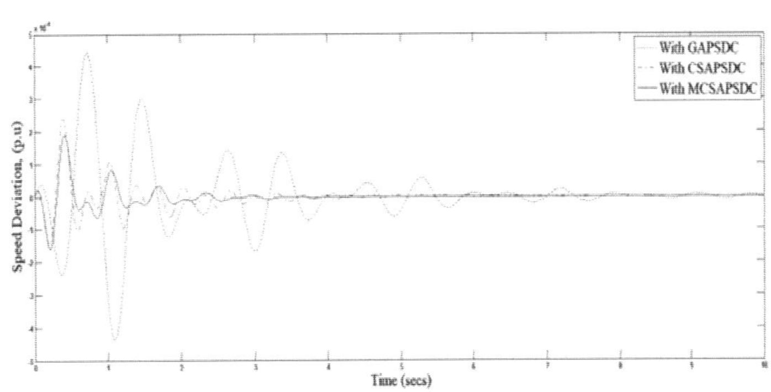

Figure 8. Speed deviation in G_2 for the condition P=0.98, Q=0.08, Pd=0.009 pu.

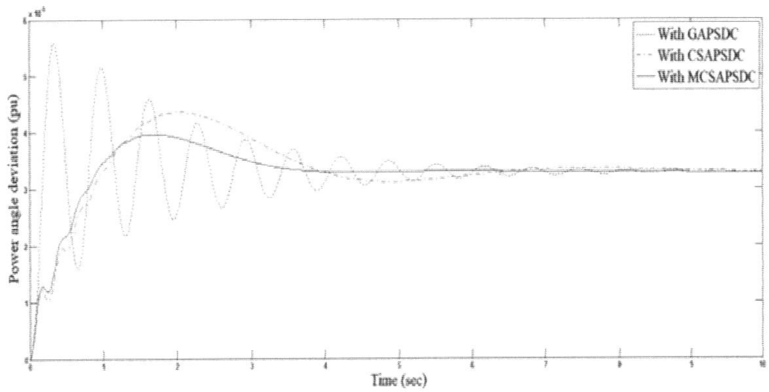

Figure 9. Power angle deviation in G_2 for the condition P=0.98, Q=0.08, Pd=0.009 pu.

Response of power angle and speed deviation of G2, G3 is shown in figure 8 and 9, with the condition of P=0.98, Q=0.08, Pd=0.009 pu, this outcome represents MCSA algorithm is effectively used for the control in compared with the GA, CSA power system damping controller. The deviation represents the chameleon- algorithm based PSDC to overcome the low frequency and make a strong system also to improve the system stability.

6. Conclusion

Multimachine system equipped with power system damping controller by using various nature inspired algorithm methods like GA, CSA and MCS. Modified CSA is implemented in three machine nine bus IEEE multimachine powers system with dual controller. Controller (PSDC) equipped in dual control system is more efficient for numerous distractions with many uncertainties. Advanced methods for solving damping ratios and error deviation can be carried out by simulations. The results of simulations are contrasted with the good damping ratios, for speed and power angle response to improve stability of the system. Modified CSA implemented in the dual input system which has great performance of damping that improves the system stability. Results are validated that modified CSA has great performance in compared with GA and conventional CSA based controllers. Proposed MCA has a better damping performance and also reduce deviations in terms of speed and power angle to improve the stability analysis with better damping ratios. In future, research work can be extended for multimachine involved more than 3 generators in system, and also more variations in operating conditions for better stability enhancement in system.

7. References

[1] Mashhadany Y Al, Abbas A K and Algburi S 2022 Study and Analysis of Power System Stability Based on FACT Controller System *Indonesian Journal of Electrical Engineering and Informatics* **10** 2

[2] Rangasamy S and Kuppusami Y 2021 A Novel Nature-Inspired Improved Grasshopper Optimization-Tuned Dual-Input Controller for Enhancing Stability of Interconnected Systems *Journal of Circuits, Systems and Computers* **30** 2150134

[3] Trivedi I N, Jangir P and Parmar S A 2016 Optimal power flow with enhancement of voltage stability and reduction of power loss using ant-lion optimizer *Cogent Eng* **3** 1208942

[4] Senjyu T, Morishima Y, Arakaki T and Uezato K 2002 Improvement of multimachine power system stability using adaptive PSS *Electric Power Components and Systems* **30** 361

[5] Butti D, Mangipudi S K and Rayapudi S R 2020 An improved whale optimization algorithm for the design of multi-machine power system stabilizer *International Transactions on Electrical Energy Systems* **30** 5

[6] Hammouri A I, Mafarja M, Al-Betar M A, Awadallah M A and Abu-Doush I 2020 An improved Dragonfly Algorithm for feature selection *Knowl Based Syst* **203** 106131

[7] Darwish A 2018 Bio-inspired computing: Algorithms review, deep analysis, and the scope of applications *Future Computing and Informatics Journal* **3** 231

[8] Butti D, Mangipudi S K and Rayapudi S 2018 Optimal Design of Modified Power System Stabilizer Using Multi Objective Based Bio Inspired Algorithms *International Journal of Energy Optimization and Engineering* **7** 210124

[9] Izadi H and Effatnejad R 2015 Power system stabilizers and control of reactive power compensator in multi-machine by particle swarm optimization algorithm *Indian J Sci Technol* **8** 254

[10] Mirjalili S 2016 Dragonfly algorithm: a new meta-heuristic optimization technique for solving single-objective, discrete, and multi-objective problems *Neural Comput Appl* **27** 1053

[11] Rangasamy S and Manickam P 2014 Stability analysis of multimachine thermal power systems using the nature-inspired modified cuckoo search algorithm *Turkish Journal of Electrical Engineering and Computer Sciences* **22** 1099

[12] Shivakumar R 2014 Implementation of an innovative cuckoo search optimizer in multimachine power system stability analysis *Control Engineering and Applied Informatics* **16** 98

[13] Tavakoli M R, Rasouli V and Allahkaram S 2015 A new design of double input Power system stabilizers using SQP for interconnected Power systems *Proceedings - International Conference on Modern Electric Power Systems, MEPS* p 16023685

[14] Kumar S, Yashaswini H K, Sharma N K, Bajaj M, Naithani D and Maindola M 2022 Classical Secondary Control Techniques in Microgrid Systems - A Review *Proceedings - 2022 2nd*

International Conference on Innovative Sustainable Computational Technologies, CISCT p 22682768

Design of H5 Transformerless Inverter for Photovoltaic System

Amrutha N Daivagna[1], Sujit Kumar[1]*, M Balamurugan[1], Mukul Mayur[1], Abhishek Kumar Tiwari[1], Akhilesh Kumar[1] and Aniket Bharti[1]

[1] Department of Electrical and Electronics Engineering, Dayananda Sagar College of Engineering, Bengaluru, Karnataka, India

sujit-eee@dayanandasagar.edu

Abstract. There are two distinct categories of sources for generating electrical power: conventional and non-conventional. At now, traditional sources such as coal, gas, and nuclear power are widely employed for the purpose of electricity generation; nevertheless, they are associated with detrimental environmental consequences. Therefore, it is crucial to utilize renewable energy sources like sun and wind to provide some of the necessary electrical power. The driving force behind this transition is the imperative for clean energy, the mitigation of greenhouse gas emissions, and the capacity to create electricity in geographically isolated regions. A proposed solution for using solar energy in single-phase AC applications involves the implementation of an H5 converter topology. The proposed architecture employs twin input DC-DC boost converters, which incorporate solar panels and batteries, to manage the desired output DC voltage. The H5 topology has notable suitability for single-phase transformerless inverters in AC applications encompassing low and high-power demands, owing to its capacity to effectively mitigate leakage current. In order to evaluate the viability of the suggested methodology, a MATLAB simulation is performed to simulate the complete procedure. Furthermore, the development of a prototype is underway in order to assess and verify the performance and outcomes of the converter. The simulation will be conducted with either the MATLAB.

1. Introduction

The use of photovoltaic (PV) energy is widely acknowledged as an exceedingly appealing and sustainable kind of renewable energy, mostly owing to its consistent availability throughout the year. According to expert projections, photovoltaic (PV) arrays are anticipated to provide a significant portion exceeding 45% of the forthcoming power requirements. The arrays have the capability to supply energy for both independent applications and utility grids [1]. The incorporation of photovoltaic (PV) arrays into utility grids has been increasingly common in recent years. The typical method employed involves the utilization of a direct current (DC) to alternating current (AC) inverter, which is subsequently followed by a line transformer. Nevertheless, conventional line transformers exhibit several limitations, including elevated system expenses, diminished efficiency, substantial dimensions, and weight [1]. Various alternative solutions have been proposed to address the problem of leakage current. One potentially viable approach to be considered entails the utilization of transformerless inverters, which function to separate the direct current (DC) component, comprising the photovoltaic (PV) array, from the alternating current (AC) component, encompassing the inverter and the grid, particularly in freewheeling modes. When it comes to transformerless systems, maximizing the power generated by the photovoltaic (PV) array is of utmost importance because this maximizes the utilization factor [2]. A maximum power point tracker (MPPT) is used to continuously identify and collect the highest possible power from the photovoltaic (PV) array, hence improving the efficiency of power extraction [3]. The proposed system incorporates a boost converter as an impedance-matching device to facilitate the connection between the photovoltaic (PV) panel and the H5 inverter. The boost converter effectively sustains the photovoltaic (PV) current at the optimal power point tracking (MPPT) value through continuous conduction mode operation. Grid-connected transformerless inverters in the H5 topology are the focus of this paper, which aims to provide a thorough analysis of their operation, regulation, and stability evaluation [4].

2. Transformerless PV inverter

Solar inverters have experienced significant adoption in business environments in recent times, owing to the emergence of solar energy as a promising candidate for the future energy supply. Despite the initially substantial investment expenditures, the cost of solar energy has experienced a notable decline in several nations as a result of government measures aimed at encouraging its use as a viable and sustainable energy alternative [3]. As a result, solar panels have gained economic feasibility, especially in scenarios characterized by a poor power grid or in remote areas without access to centralized electricity distribution. In this regard, the use of voltage source inverters (VSI) in photovoltaic (PV) inverters has been seen in figure 1.

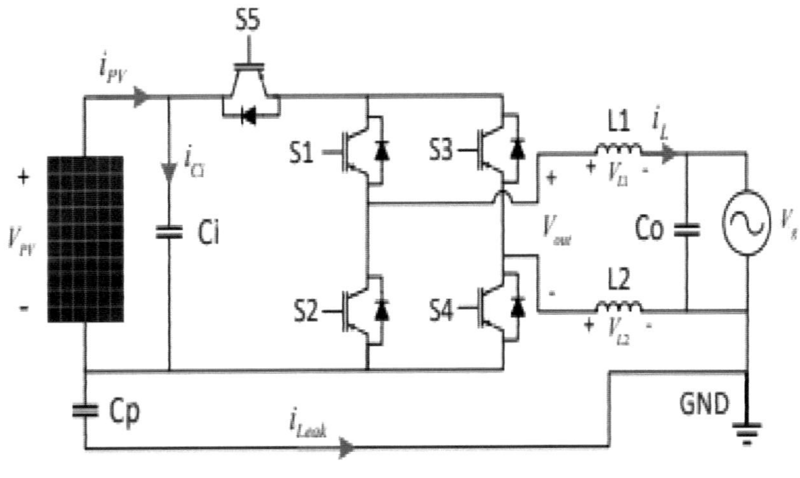

Figure 1. The H5 transformerless inverter exhibiting a typical arrangement.

The progressive evolution of maximum power point tracking (MPPT) methodologies has played a significant role in the advancement and enhanced efficacy of solar photovoltaic (PV) systems. The technological progressions have resulted in improvements in the dependability, effectiveness, and steadfastness. Considerable resources have been allocated towards the enhancement of solar photovoltaic (PV) systems via extensive research and development endeavors [4].

The representation of the PV single-phase transformerless inverter system may be depicted using a fundamental block diagram. The affordability of solar inverters has exerted a substantial influence on the proliferation of photovoltaic systems. The incorporation of galvanic isolation, which ensures separation between the panel and the load end, may lead to increased system size and reduced efficiency owing to losses in the core and windings. The total efficiency of the DC-DC converter is impacted by the combination of leakage and a high-frequency transformer [5].

Nevertheless, the utilization of transformerless technology presents significant benefits, including as a reduction in dimensions, expenses, and mass, alongside an overall enhancement in the efficiency of the system. Addressing specific safety concerns pertaining to the parasitic capacitance of the solar panel has significant importance. The leakage capacitance is subject to several influencing elements, including but not limited to the spatial separation between the module, the level of humidity, the presence of dust cover, and the prevailing weather conditions at the site of installation [6].

3. Proposed methodology

The PV-grid-connected system being examined consists of two separate stages. The first regulation phase is distinguished by deploying the Maximum PowerPoint (MPP) controller. This methodology enables the optimization of power extraction from the photovoltaic (PV) panel, maximizing its output capacity [7,8]. The optimization of the photovoltaic (PV) panel's performance is achieved by the implementation of a duty cycle, which ensures that the panel functions within the conditions of maximum power point tracking (MPPT). The regulated constraints in this stage include the voltage of the DC connection and the current flowing through the grid. Figure 2 illustrates the presence of two vertically aligned loops within the period. The internal loop is called the grid current, while the outside is characterized as the DC link voltage.

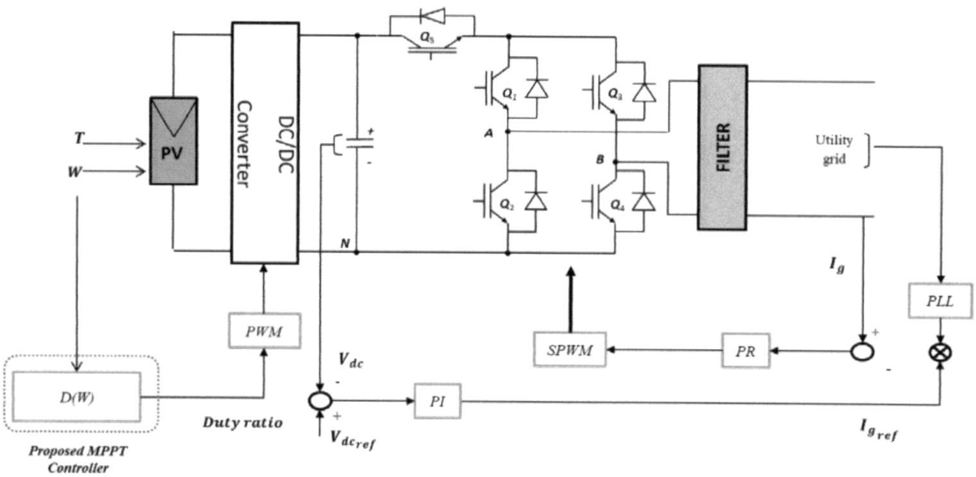

Figure 2. Proposed system with controllers.

3.1. Proposed MPPT controller

The escalating need for electricity and the growing recognition of environmental concerns has driven the proliferation of photovoltaic (PV) systems. The fluctuation in sun irradiation has posed difficulties in optimizing power extraction. Therefore, numerous methodologies have been developed for Maximum PowerPoint Tracking (MPPT) [9–12]. There are three primary types of MPP approaches: offline, online, and hybrid. This research delves into the offline technique of producing the locus signal (short circuit current, open circuit voltage, solar irradiance, and temperature) required to monitor the MPP [13–16]. FOCV and FCCC are extensively used offline methods to isolate the PV array and set the MPP process operating point. These methods are simple; however, they cause occasional power loss owing to PV module separation.

The FCCC [16] is based on the premise that the Maximum Power Point (MPP) current may be estimated by multiplying the short circuit current with a factor referred to as "k". A standard range for "k" is between 0.71 and 0.90, determined by referring to the PV array's datasheet [17–24]. One of the primary concerns about the FCCC is to the challenge of accurately quantifying the short circuit current during the operation of the photovoltaic (PV) system [25]. The Fraction of Open Circuit Voltage (FOCV) is determined by the linear relationship between the PV array voltage at Maximum Power Point and the open circuit voltage. This relationship is governed by a factor, denoted as k, which is specified in the PV array datasheet and typically ranges between 0.71 and 0.80 [26].

Using a curve-fitting approach, A mathematical model for MPP may be generated offline using the MPPT approach, which uses panel characteristics curves [26]. In this study, the MPP is derived from the P-V curve. The P-V curves for various levels of irradiance are generated using the PV panel's datasheet, as illustrated in figure 3.

The input of the controller is the insolation level measured in watts. At the same time, its output is the duty value, represented as d(t) and depicted in figure 7. The H5 controller regulates the boost converter's output to maintain a constant voltage of VDC (VDC = 450 V). Given the assumption of continuous conduction in the boost converter, the relationship between the input and output voltages may be found in reference [27]. To achieve MPPT conditions, the terminal voltage of the photovoltaic (PV) system must be equivalent to Vmpp, which represents the voltage at the maximum power point (MPP). The duty cycle during full power point circumstances, denoted as dmpp, as established by reference [28]. The maximum power point (Vmpp) is determined by analysing the photovoltaic distinctive curves associated with different levels of insolation, as seen in figure 3.

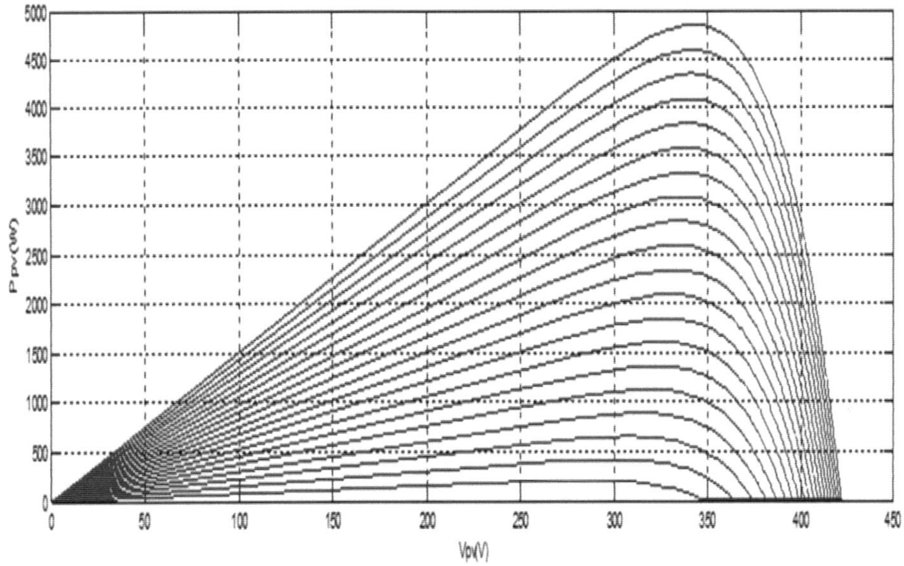

Figure 3. P-V characteristics curves.

3.2. Proposed controller H5

The controller possesses dual purposes. The primary aim is to ensure the voltage, denoted as VDC, remains consistently stable. The second objective is to mandate grid current tracking to reach an established target value. The controller has two loops: an inner loop that controls the grid current and an outside loop that controls the VDC. To maintain stability, it is generally necessary for the internal control loop to exhibit a faster response time compared to the outer loop [27,28].

The presence of a high-value capacitor at the input side of H5 in the outer loop results in a reduced speed of response. Implementing a basic proportional-integral (PI) controller has been employed for the control loop in question. The proportional integral gains, commonly called PI, are tweaked using the Nichehols approach.

The inner loop uses a sinusoidal waveform as its reference signal alongside a grid current reference. The outer loop controller controls the wave's amplitude, while the phase is obtained from a phase-locked loop (PLL) that is time-locked to the power grid. Inverters linked to the grid usually use a Proportional Resonant (PR) controller [26,28]. The controller mentioned above demonstrates significant utility when employed with sinusoidal reference signals since they yield a response that is deemed more satisfactory in comparison to that of PI controllers.

4. Result and discussion

H5 inverters are quite like single-phase full-bridge inverters structurally, with the accumulation of a DC-bypass switch marked by "S5". When there is a continuous flow of power with no impediment, this switch disconnects the photovoltaic (PV) array from the utility grid. Figure 1 shows a graphic representation of the H5 inverter design, highlighting the presence of leakage current (ILeak) amid the PV array and ground.

In figure 4, most H5 inverters offer four distinct operating modes. Figure 4(a) depicts the active mode, which can be observed in operation throughout the positive half-cycle. The second working mode, seen in figure 4(b), includes the zero-voltage vector and is usually referred to as the current-freewheeling mode. S1 is activated while S4 and S5 are deactivated in the scenario. S3 acts as a diode that allows current to flow freely.

The H5 inverter is third mode of operation corresponds to an active state that exists only during the negative half-cycle, as shown in figure 4(c). S1 switch behaves as a freewheeling diode, like the configuration of S3 in mode 2 (figure 4(b)). Space vector modulation (SVM) is used by the H5 inverter, and its modes of operation and implementation are shown in table 1 and figure 2, respectively

Both the existing transformer-based H5 inverter system and the proposed transformer-free H5 inverter system connected to the PV array depicted in figure 1 and figure 4 are simulated in Matlab/Simulink. Table 2 displays the measured values of the system's parameters. Both the traditional and proposed methods use photovoltaic (PV) arrays with the same total number of PV modules. However, the number of PV modules in parallel and series needed to maintain the direct current (DC) bus voltage (VDC) varies. The traditional form's photovoltaic (PV) panel configuration consists of 910 series cells in 2 parallel strings, whereas the proposed configuration involves 455 series cells in 4 similar strings. The overall sales for both layouts are equal, amounting to 1820 cells. The switching frequency of the boost converter was 5 kHz.

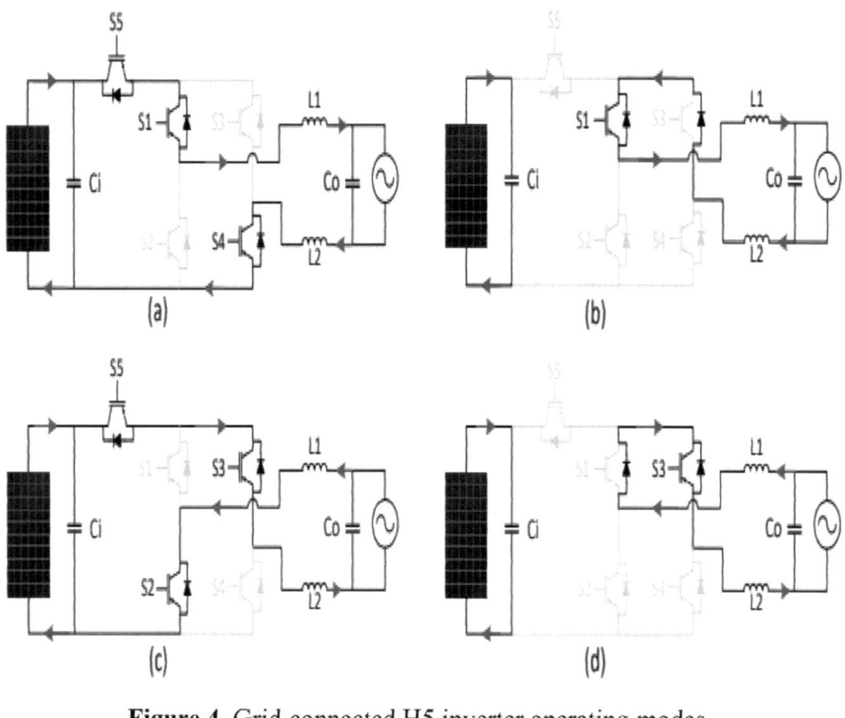

Figure 4. Grid-connected H5 inverter operating modes.

Table 1. H5 inverter switching states.

Mode	S1	S2	S3	S4	S5	V_{out}

1	1	0	0	1	1	V_{PV}
2	1	0	0	0	0	0
3	0	1	1	0	1	$-V_{PV}$
4	0	0	1	0	0	0

Table 2. PV parameters

System	Parameter	Values
	Isc Current	11.8
	Voc Voltage	563
Proposed	PV Isc Current	15.35
	PV Voc Voltage	423 V

Figure 5 shows the voltage and current waveforms from the grid, the PV system, and the leakage current of standard H5. Both cases necessitate the implementation of grid current control to confirm that it remains in-phase with voltage grid, which in turn makes it possible to maintain a power factor of unity. Moreover, it is worth noting that the earth leakage currents exhibit a virtually identical root mean square (RMS) current of 18 mA, a value that falls within the range of standard suggested values as shown by reference [27]. The configuration of the system places a significant emphasis on the MPPT operation.

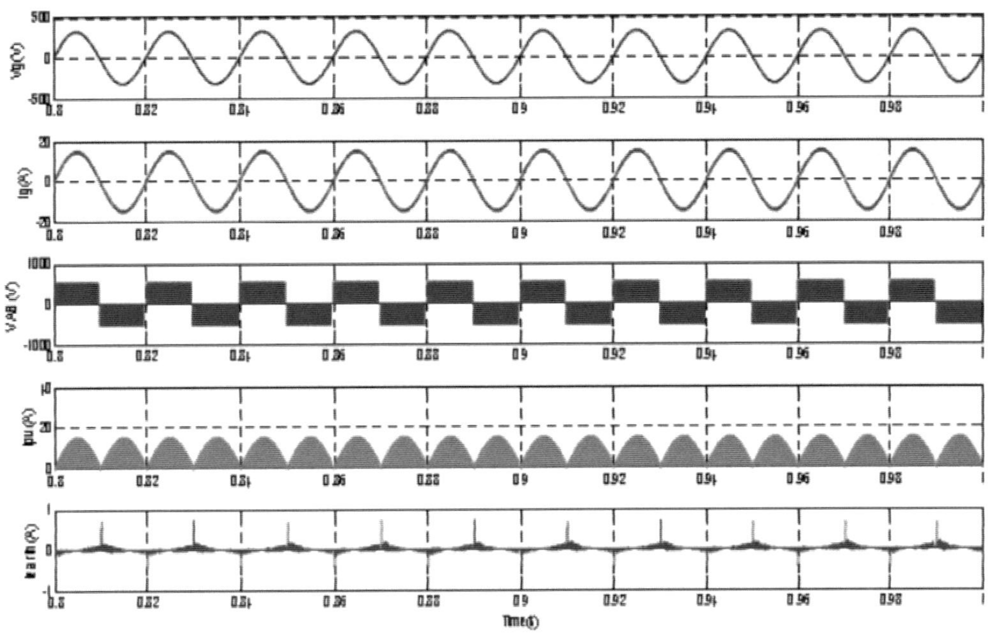

Figure 5. Comparison in terms of voltage, current (grid), current (PV) and leakage current for the standard H5 inverter.

IOP Conf. Series: Materials Science and Engineering 1295 (2023) 012016 doi:10.1088/1757-899X/1295/1/012016

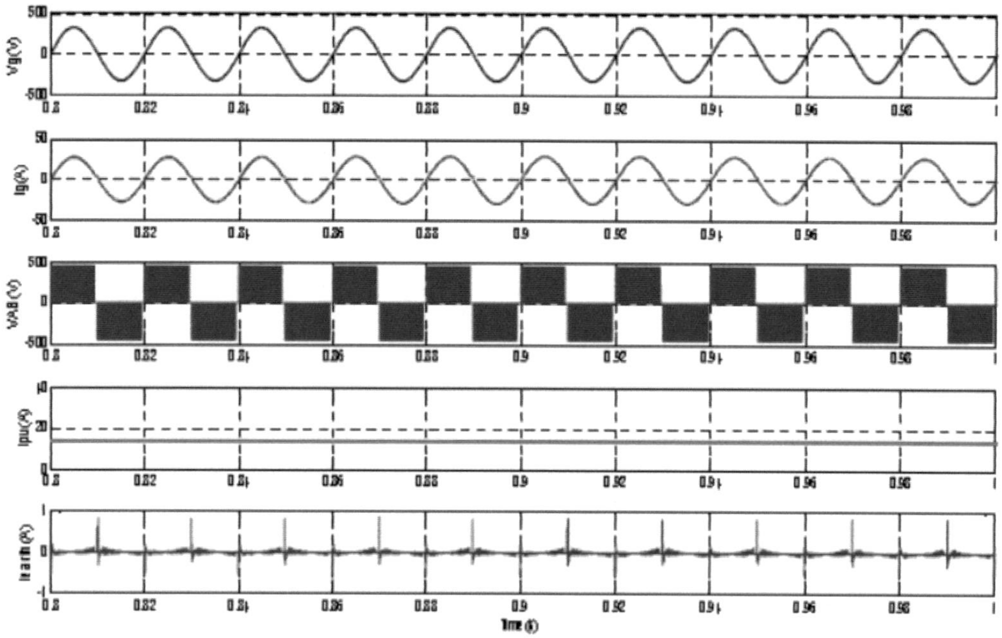

Figure 6. Comparison in terms of voltage, current (grid), current (PV) and leakage current for the H5 proposed inverter.

Figure 7 shows how the efficiency of the proposed and traditional systems varies throughout a range of output powers. Both methods were found to be relatively effective. As a result of the proposed system's additional boost converter losses, its overall efficiency is slightly lower than that of the standard system. The maximum efficiency is close to 50% of the nominal output power.

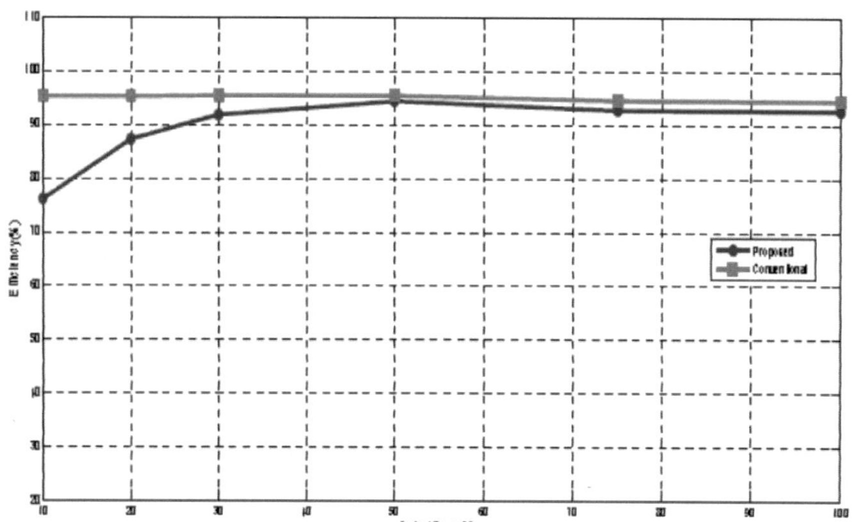

Figure 7. The efficiency of both existing and proposed systems varies with their output power.

Figure 8 depicts the comparative representation of the proposed and conventional transformerless H5 systems. The recommended system frequently demonstrates a reduced magnitude of total harmonic distortion (THD) in comparison to the standard technique. The systems under consideration offer a reliable and efficient means of supplying the utility grid with superior-quality power and free from impurities. At lower power levels (specifically, 20% of the maximum capacity), there is a little increase in total harmonic distortion (THD). If the injected power is minimal, the minor increase does not pose a problem.

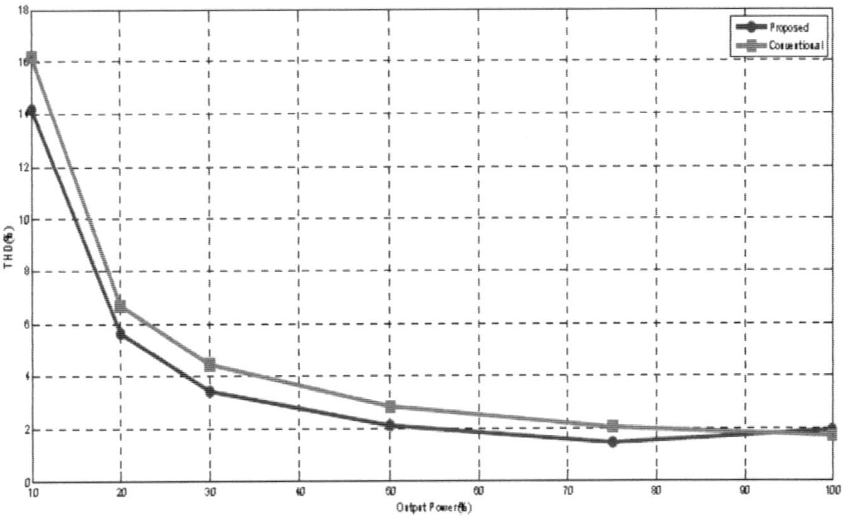

Figure 8. THD fluctuations with output power of conventional and proposed systems.

Figure 9 displays the harmonic spectrums of the grid currents for both systems. The sequential arrangement of harmonics is observed in both spectra, encompassing order and current harmonics. Notably, the suggested system exhibits relatively diminished current harmonics compared to the harmonics exhibited in a conventional system. In general, the total harmonic distortion (THD) of the suggested system exhibits a reduced magnitude when compared to the THD observed in the traditional system.

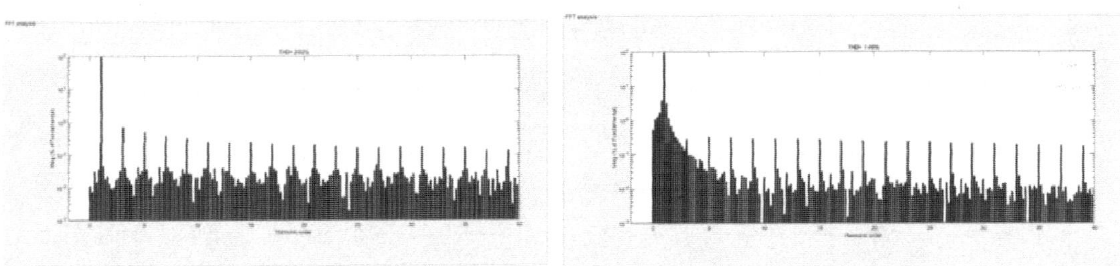

Figure 9. Typical and alternative harmonic spectra at 75% insolation.

The comparative analysis is conducted between the suggested MPPT algorithm and the conventional MPPT technique employing the "Incremental Conductance (IC)" approach. Step variations in insolation are depicted in figure 10 for the proposed system's MPPT. The relationship between the output power and the maximum power point tracking (MPPT) power is observed, wherein a slight steady state inaccuracy is present, indicating losses inside the circuit. The system's huge capacitors and the utilization of an open-loop MPPT control approach contribute to the observed comparatively delayed power response. The controller modifies the duty ratio (d) to ensure that the boost converter and photovoltaic (PV) system work under maximum power point tracking (MPPT) conditions. Changes in insolation, from 10% to 100%, cause small shifts in the duty ratio, but only within a narrow range of 22% to 35%. Changes in "MPP power, output power, and boost converter duty ratio (d)" in response to sudden variations in insolation are shown for the conventional MPPT control system in figure 9. The power output responses depicted in figures 10 and 11 exhibit similarities, although it is noteworthy that the suggested controller has a high degree of simplicity.

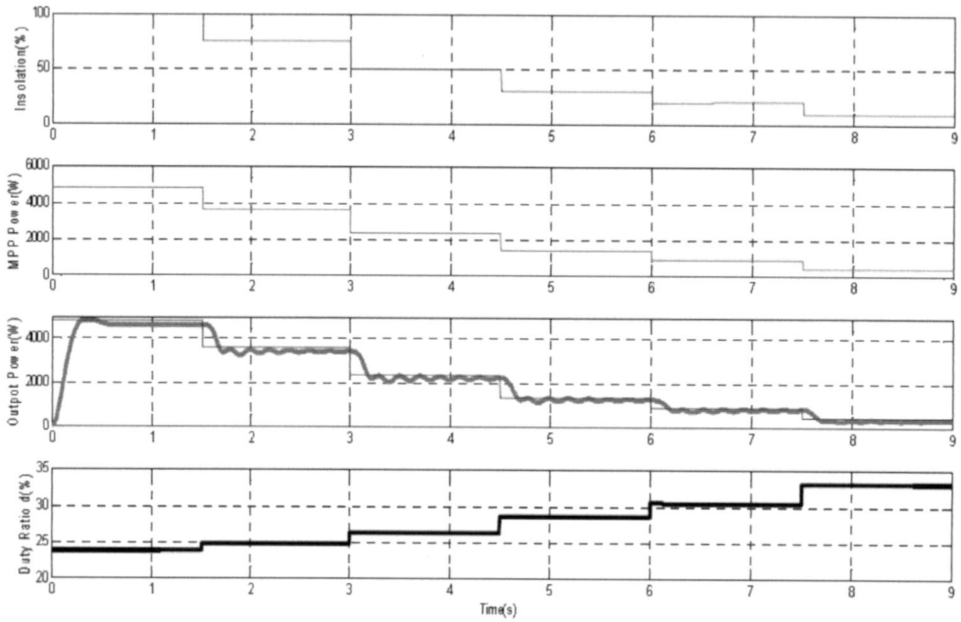

Figure 10. The proposed system's output MPP power, Pout, and DR fluctuate with step insolation.

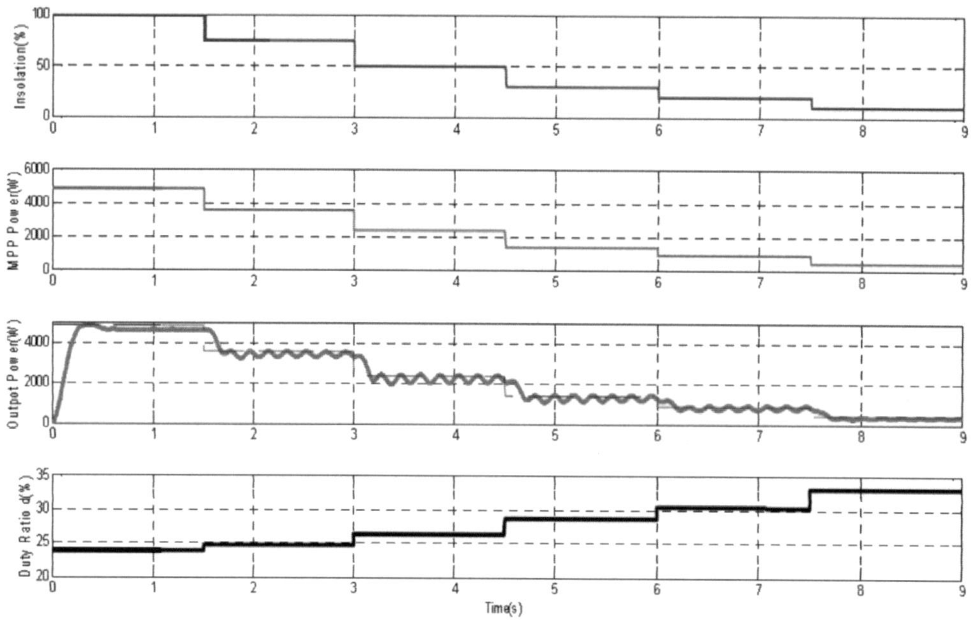

Figure 11. Conventional MPPT system with output MPP power, Pout, and DR alter at step insolation.

5. Conclusion

This research shows how to improve the H5 transformerless inverter's performance when fed energy from photovoltaic cells. The proposed architecture improves the maximum power point operation, among other performance features. Furthermore, this paper presents a basic offline Maximum Power Point Tracking (MPPT) technique. The simulation findings indicate that the utilization factor of the suggested system is significantly higher, approaching 100%, compared to the conventional system. The efficiency of the suggested system exhibits a slight decrease compared to the decline observed in the traditional method. The study also examines the photovoltaic (PV) system's performance in the suggested configuration compared to the conventional system under maximum power point tracking (MPPT) circumstances. The comparative analysis has revealed that the proposed Maximum Power Point Tracking (MPPT) algorithm exhibits a higher degree of simplicity than the conventional method.

6. References

[1] Mehrasa M, Pouresmaeil E, Pournazarian B, Sepehr A, Marzband M and Catalão J P S 2018 Synchronous resonant control technique to address power grid instability problems due to high renewables penetration *Energies (Basel)* **11** 2469

[2] Mehrasa M, Pouresmaeil E, Taheri S, Vechiu I and Catalao J P S 2018 Novel Control Strategy for Modular Multilevel Converters Based on Differential Flatness Theory *IEEE J Emerg Sel Top Power Electron* **6** 888

[3] Mehrasa M, Pouresmaeil E, Zabihi S, Vechiu I and Catalão J P S 2018 A multi-loop control technique for the stable operation of modular multilevel converters in HVDC transmission systems *International Journal of Electrical Power and Energy Systems* **96** 194

[4] Rouzbehi K, Davarifar M, Martino M, Citro C, Luna A, Daneshifar Z and Rodriguez P 2016 Comparative efficiency study of single phase photovoltaic grid connected inverters using PLECS® *2nd International Congress on Technology, Communication and Knowledge, ICTCK* p 16358208

[5] Li H, Zeng Y, Zheng T Q and Zhang B 2016 A novel H5-D topology for transformerless photovoltaic grid-connected inverter application *2016 IEEE 8th International Power Electronics and Motion Control Conference, IPEMC-ECCE Asia* p 16155197

[6] Freddy T K S, Rahim N A, Hew W P and Che H S 2015 Modulation techniques to reduce leakage current in three-phase transformerless H7 photovoltaic inverter *IEEE Transactions on Industrial Electronics* **62** 322

[7] Li H, Zeng Y, Zhang B, Zheng T Q, Hao R and Yang Z 2019 An Improved H5 Topology with Low Common-Mode Current for Transformerless PV Grid-Connected Inverter *IEEE Trans Power Electron* **34** 1254

[8] Zaid S A and Kassem A M 2017 Review, analysis and improving the utilization factor of a PV-grid connected system via HERIC transformerless approach *Renewable and Sustainable Energy Reviews* **73** 1061

[9] Calais M, Myrzik J, Spooner T and Agelidis V G 2002 Inverters for single-phase grid connected photovoltaic systems - An overview *PESC Record - IEEE Annual Power Electronics Specialists Conference* p 7440147

[10] Tan Kheng Suan F, Rahim N A and Hew W P 2011 Modeling, analysis and control of various types of transformerless grid connected PV inverters *2011 IEEE 1st Conference on Clean Energy and Technology, CET* p 12304847

[11] Dhara S and Somasekhar V T 2022 A Three-Phase Semi-Single Stage PV Inverter with Voltage Boosting and Leakage Current Minimization *IEEE Transactions on Circuits and Systems II: Express Briefs* **69** 169

[12] Mehrasa M, Pouresmaeil E, Akorede M F, Zabihi S and Catalão J P S 2017 Function-based modulation control for modular multilevel converters under varying loading and parameters conditions *IET Generation, Transmission and Distribution* **11** 3222

[13] Pouresmaeil E, Mehrasa M and Catalão J P S 2014 Control strategy for the stable operation of multilevel converter topologies in DG technology *Proceedings - 2014 Power Systems Computation Conference, PSCC* p 14919893

[14] Blaabjerg F, Teodorescu R, Liserre M and Timbus A V. 2006 Overview of control and grid synchronization for distributed power generation systems *IEEE Transactions on Industrial Electronics* **53** 1398

[15] Xiao H and Xie S 2010 Leakage current analytical model and application in single-phase transformerless photovoltaic grid-connected inverter *IEEE Trans Electromagn Compat* **52** 902

[16] López Ó, Freijedo F D, Yepes A G, Fernández-Comesaña P, Malvar J, Doval-Gandoy J and Teodorescu R 2010 Eliminating ground current in a transformerless photovoltaic application *IEEE Transactions on Energy Conversion* **25** 140

[17] Siwakoti Y P and Blaabjerg F 2018 Common-ground-type transformerless inverters for single-phase solar photovoltaic systems *IEEE Transactions on Industrial Electronics* **65** 2100

[18] González R, Gubía E, López J and Marroyo L 2008 Transformerless single-phase multilevel-based photovoltaic inverter *IEEE Transactions on Industrial Electronics* **55** 2694

[19] Hase Y 2012 *Handbook of Power Systems Engineering with Power Electronics Applications*

[20] Zhang L, Sun K, Xing Y and Xing M 2014 H6 transformerless full-bridge PV grid-tied inverters *IEEE Trans Power Electron* **29** 1229

[21] Rizzoli G, Mengoni M, Zarri L, Tani A, Serra G and Casadei D 2016 Comparison of single-phase H4, H5, H6 inverters for transformerless photovoltaic applications *IECON Proceedings (Industrial Electronics Conference)* p 16557339

[22] Guo X, Jia X, Lu Z and Guerrero J M 2016 Single phase cascaded H5 inverter with leakage current elimination for transformerless photovoltaic system *Conference Proceedings - IEEE Applied Power Electronics Conference and Exposition - APEC* p 15987772

[23] Gotekar P S, Muley S P, Kothari D P and Umre B S 2016 Comparison of full bridge bipolar, H5, H6 and HERIC inverter for single phase photovoltaic systems - A review *12th IEEE International Conference Electronics, Energy, Environment, Communication, Computer, Control: (E3-C3), INDICON* p 15888156

[24] Araújo S V, Zacharias P and Mallwitz R 2010 Highly efficient single-phase transformerless inverters for grid-connected photovoltaic systems *IEEE Transactions on Industrial Electronics* **57** 3118

[25] Li W, Gu Y, Luo H, Cui W, He X and Xia C 2015 Topology review and derivation methodology of single-phase transformerless photovoltaic inverters for leakage current suppression *IEEE Transactions on Industrial Electronics* **62** 4537

[26] Kassem A M 2012 MPPT control design and performance improvements of a PV generator powered DC motor-pump system based on artificial neural networks *International Journal of Electrical Power and Energy Systems* **43** 90

[27] Asim M, Tariq M, Mallick M A, Ashraf I, Kumari S and Bhoi A K 2018 Critical evaluation of offline MPPT techniques of solar PV for stand-alone applications *Lecture Notes in Electrical Engineering* vol 435 p 13–21

[28] Sher H A, Murtaza A F, Addoweesh K E and Chiaberge M 2013 An intelligent off-line MPPT technique for PV applications *Proceedings - 2013 IEEE Conference on Systems, Process and Control, ICSPC* p 14096104

IOP Conf. Series: Materials Science and Engineering 1295 (2023) 012017 doi:10.1088/1757-899X/1295/1/012017

Optimal Power Flow Management in Microgrids using Distributed Energy Resources

Nagananda R[1*], and Gopiya Naik S[1]

[1] Department of Electrical and Electronics Engineering, P.E.S College of Engineering, Mandya, Karnataka, India

nagananda22@gmail.com

Abstract. The effective and stable functioning of microgrids using distributed energy resources (DERs) is greatly dependent on ideal power stream management. To get the optimal power flow in microgrids, this study provides an inquiry into the use of cutting-edge optimization techniques. To fulfil load demand and consider DER requirements, system running costs must be kept to a minimum. The microgrid integrates a variety of distributed energy resources, including solar photovoltaic systems, wind turbines, and energy storage systems. The research examines several optimization strategies, such as mixed-integer programming, evolutionary algorithms, and linear programming, to decide on the best DER dispatch and scheduling. The findings give important information on how to efficiently use and manage distributed energy resources to raise the performance and resilience of microgrids.

1. Introduction

The modern power grid is undergoing a transformation due to various factors, such as increasing demand for electricity, the integration of renewable energy sources, and the emergence of new technologies [1]. These changes have led to the rise of microgrids as a viable solution to address the challenges associated with the traditional centralized power generation and distribution system. Microgrids are localized energy systems that can operate independently or in coordination with the main grid. They consist of Distributed Energy Resources (DERs), including solar PV, wind turbines, microturbines, fuel cells, and energy storage systems, among others. The integration of these DERs offers numerous benefits, such as enhanced energy efficiency, grid resilience, and reduced greenhouse gas emissions [2].

The primary objective of this research paper is to investigate and evaluate the implementation of Optimal Power Flow (OPF) techniques in microgrids, focusing on the effective management of distributed energy resources. The research aims to achieve the following goals [3]:

- To assess the status of microgrids and distributed energy resources, highlighting integration challenges and opportunities.
- To analyse existing OPF techniques and algorithms in microgrids, considering their advantages and limitations.
- To propose an innovative and efficient OPF management framework in microgrids that optimally utilizes various distributed energy resources while ensuring grid stability and reliability.
- To conduct case studies and simulations to validate the proposed framework's performance and compare it with traditional control systems.

This research work will focus on microgrid systems that incorporate various types of distributed energy resources, such as solar PV, wind turbines, energy storage systems (ESS), and small-scale generators

[4]. The study will aim to optimize the power flow within the microgrid to achieve operational objectives, considering technical constraints, economic factors, and environmental sustainability. While the research will emphasize the use of OPF methods, other control techniques may be explored for comparative analysis.

By investigating optimal power flow management in microgrids with distributed energy resources, this research aims to provide valuable insights into enhancing the efficiency, reliability, and sustainability of microgrid operations [5]. These findings will play a pivotal role in advancing the development and deployment of microgrid technologies, facilitating the integration of renewable energy sources, and supporting the transition towards a more resilient and sustainable energy future.

2. System configuration

The concept of microgrids has garnered significant attention in recent years as a viable solution to enhance energy resilience and reliability [6]. Microgrids are characterized by their ability to operate in grid-connected or islanded mode, providing continuous power supply during grid blackouts. Various control methodologies have been developed to ensure seamless microgrid operation. Some of the key control approaches include:

- Centralized Control: In this method, a central controller manages the operation of distributed energy resources and directs power flow based on real-time data. Centralized control offers effective optimization but can be susceptible to single points of failure [7].
- Distributed Control: In contrast to centralized control, this approach distributes control functions across multiple nodes within the microgrid. Distributed control enhances system resilience and flexibility but may require more advanced communication and coordination networks.
- Hierarchical Control: Hierarchical control combines elements of both centralized and distributed control. It involves multiple control levels, with local controllers overseeing individual DERs, while a higher-level controller coordinates overall system operation.

2.1. Optimal power flow in microgrids

Optimal Power Flow (OPF) is a robust optimization technique used to determine the optimal power setpoints of generators and controllable loads within a power system. Its purpose is to minimize operational costs or achieve other performance objectives while satisfying operational constraints [8]. OPF has been successfully applied in conventional power systems and is gaining popularity in microgrid management.

- Linear Programming (LP)-based OPF: LP-based OPF is one of the earliest methods used for microgrid optimization. It formulates the power flow problem as a linear program, representing objective functions and constraints as linear equations. While LP is computationally efficient, it may not adequately handle nonlinearities and complex constraints in microgrid systems.
- Non-Linear Programming (NLP)-based OPF: NLP-based techniques consider nonlinear equations and constraints, enabling more accurate modelling of microgrid components. However, they can be computationally intensive, especially for large-scale microgrids.
- Evolutionary Algorithms: Evolutionary algorithms, such as genetic algorithms and particle swarm optimization, provide a heuristic approach to solving OPF problems. They are particularly suitable for multi-objective optimization and can efficiently handle both linear and nonlinear optimization problems.

2.2. Reconciliation of conveyed energy assets in microgrids

The integration of different DERs in microgrids presents several challenges due to the intermittent nature of renewable energy sources and the complexity of managing multiple energy sources simultaneously [9,10].

- Solar PV Coordination: Solar PV systems are widely used in microgrids due to their availability and environmental benefits. Effective coordination requires accurate forecasting methods to anticipate solar output fluctuations.

- Wind Power Integration: Wind turbines can be integrated into microgrids to complement solar PV systems. Dynamic control strategies are necessary to handle the variability and uncertainty of wind power generation.
- Energy Storage Systems (ESS): ESS play a crucial role in smoothing fluctuations in renewable energy sources and enhancing microgrid stability. Various ESS technologies, such as batteries and flywheels, have been explored for optimal operation in microgrids.
- Hybrid Microgrids: Hybrid microgrids combine multiple DERs to ensure reliable and continuous energy supply. The proper integration of various components is essential for achieving optimal performance.

2.3. Existing exploration on OPF in microgrids

A Notwithstanding the advancements in microgrid OPF research, a few challenges and valuable opportunities remain:

- Versatility: Many existing OPF approaches may not scale well for large and complex microgrids. Developing flexible algorithms is essential to accommodate future expansion.
- Integration of Emerging Technologies: As new DERs and control technologies emerge, their integration and optimization within microgrids require further investigation.
- Consideration of Secondary Services: Microgrids may also provide secondary services to the main grid. Integrating these services into OPF development can enhance the microgrid's economic viability.
- Interoperability and Standardization: Establishing standards and protocols for microgrid control and communication will facilitate seamless integration and interoperability among different microgrid components.

The literature review reveals the significance of optimal power flow management in microgrids with distributed energy resources [11]. Various control strategies and optimization techniques have been proposed and evaluated to enhance microgrid performance. The upcoming sections will build upon this knowledge to present an innovative framework for optimal power flow management in microgrids, addressing the identified gaps and opportunities.

3. Overview of the optimal power flow management system

To achieve the goals of this study, an efficient strategy is employed for the analysis and development of an optimal power flow management system for microgrids with distributed energy resources. The process comprises the following key steps [12]:

Step -1 Microgrid Framework Model: The initial step involves creating a comprehensive model of the microgrid system, considering various distributed energy resources and their characteristics. The components to be included in the model may consist of solar PV systems, wind turbines, energy storage systems (ESS), small-scale generators, and controllable loads. Mathematical representations and data for each component, such as power output curves, efficiency profiles, state of charge (SOC) characteristics for ESS, and load demand profiles, are collected.

Step -2 Objective Function Formulation: The next stage is to formulate the objective function that will be optimized during the OPF process. The objective function may be designed to minimize operational costs, maximize renewable energy utilization, reduce fossil fuel emissions, or achieve a combination of these objectives. It will be a function of the control variables, representing the setpoints of generators, ESS, and loads, and may include factors such as fuel costs, emissions coefficients, and renewable energy availability.

Step -3 Optimization Algorithm Selection: With the system model and objective function in place, an appropriate optimization algorithm is chosen to solve the microgrid's OPF problem efficiently [13]. The choice of the algorithm will depend on factors such as the complexity of the microgrid, the size of the optimization problem, and the desired solution accuracy. Commonly used algorithms for microgrid OPF include Linear Programming (LP), Non-Linear Programming (NLP), and Evolutionary Algorithms like Genetic Algorithms or Particle Swarm Optimization.

Step -4 Constraints and Operational Requirements: To ensure the microgrid's stable and reliable operation, various constraints and operational requirements must be considered during the development process. These constraints may include power balance conditions, voltage limits, current limits, ramp rate limits, and SOC constraints for energy storage systems. Additionally, any regulatory or contractual requirements, such as power purchase agreements or grid codes, should be accounted for.
Step -5 Proposed OPF Management Framework: Based on the selected optimization algorithm and considering the constraints and objectives, the research proposes an innovative OPF management framework tailored to microgrid systems with distributed energy resources. Figure 1 indicates the proposed framework will integrate the specific features and dynamics of microgrid operation to achieve enhanced efficiency, resilience, and sustainability.

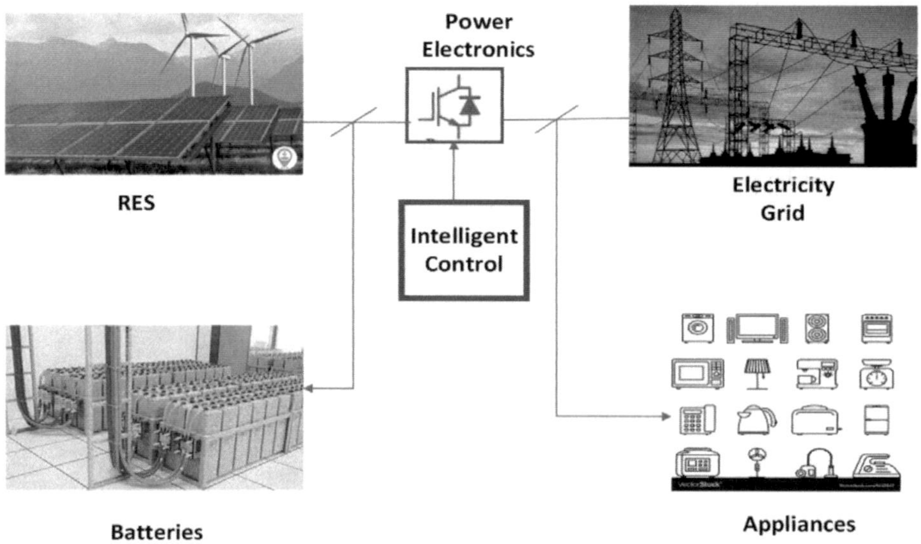

Figure 1. Configuration of RESs and power electronics with intelligent control.

Step -6 Case Studies and Simulations: To validate the effectiveness of the proposed framework, case studies and simulations will be conducted using representative microgrid scenarios. These scenarios will encompass various operating conditions, including different renewable energy availability, load patterns, and unexpected disturbances. The performance of the proposed OPF management framework will be compared with existing control strategies, such as centralized control, distributed control, and hierarchical control, to assess its superiority.
Step -7 Performance Evaluation Metrics: Performance evaluation metrics will be defined to quantitatively measure the effectiveness of the proposed OPF management framework. Key performance indicators may include the percentage of renewable energy penetration, total operational cost savings, reduction in fossil fuel emissions, grid stability indices, and the ability to meet load demand under various scenarios.
Step -8 Data Collection and Software Tools: The research relies on data collection from real microgrid installations or realistic synthetic datasets to ensure the validity and accuracy of the simulations. Additionally, software tools and simulation platforms, such as MATLAB, GAMS, or specialized microgrid simulation tools, will be used to implement and analyse the proposed OPF management framework [14].
Step -9 Sensitivity Analysis and Robustness: A sensitivity analysis will be conducted to investigate the impact of uncertainties, such as variations in renewable energy generation and load demand forecasting errors, on the performance of the proposed OPF system. The research will explore the robustness of the system under different scenarios and assess its ability to adapt to dynamic microgrid conditions.

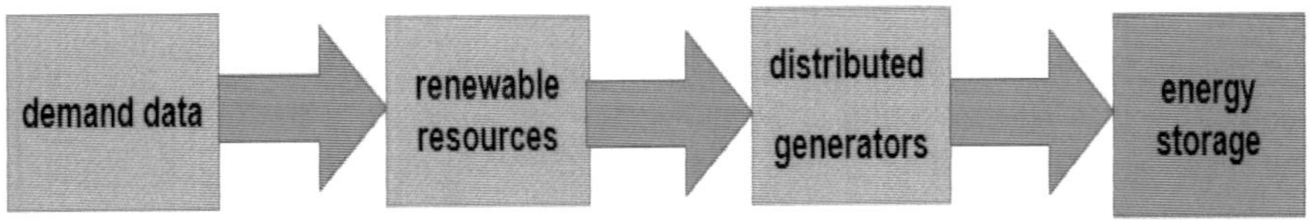

Figure 2. Methods followed by integrating distributed energy resources into microgrids.

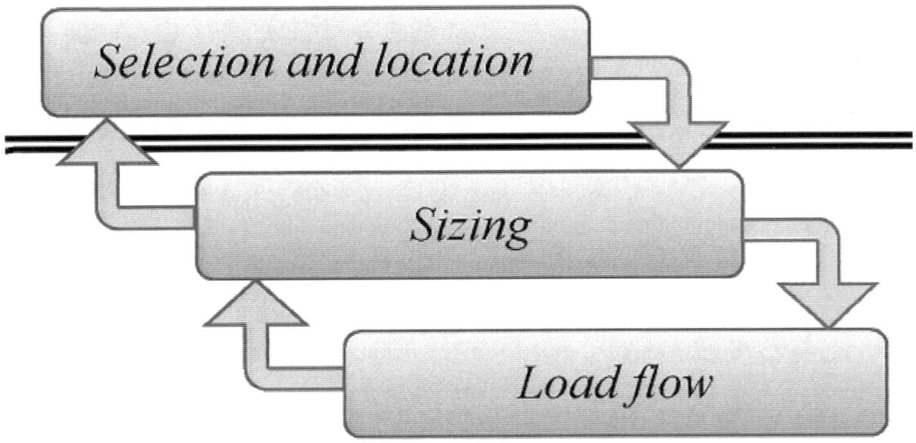

Figure 3. Methods for an integration of DGs and ESSs.

The above methodology ensures a systematic and comprehensive approach to develop and evaluate the proposed optimal power flow management system for microgrids with distributed energy resources [15]. It addresses the specific challenges of microgrid operation, considers the integration of various DERs, and aims to enhance microgrid performance based on identified objectives and constraints.

4. Methodology

To achieve the objectives of this study, an efficient strategy is adopted for the analysis and enhancement of an optimal power flow management system for microgrids with distributed energy resources (DERs). The initial step involves outlining a comprehensive model of the microgrid system, considering various distributed energy resources and their characteristics. Components to be included in the model may consist of solar PV systems, wind turbines, energy storage systems (ESS), small-scale generators, and controllable loads. Numerical representations and data for each component, such as power output curves, efficiency profiles, state-of-charge (SOC) values for ESS, and load demand profiles, are gathered.
The next step is to formulate the objective function that will be optimized during the OPF process. The objective function may be designed to minimize operational costs, maximize renewable energy utilization, reduce fossil fuel emissions, or achieve a combination of these goals. It will be a function of the control variables, representing the setpoints of generators, ESS, and loads, and may include variables such as fuel costs, emissions coefficients, and renewable energy availability.
With the system model and objective function in place, an appropriate optimization algorithm is selected to solve the microgrid's OPF problem efficiently [13]. The preferred algorithm would depend on various factors such as the optimization problem size, level of accuracy desired, and microgrid intricacy. Some algorithms commonly used in microgrid OPF are Linear Programming (LP), Non-Linear Programming (NLP), and Evolutionary Algorithms such as Genetic Algorithms and Particle Swarm Optimization.
During microgrid development, factors like power balance, voltage limits, and SOC constraints must be considered for reliable and stable operation. Other operational demands like current limits and ramp rate

limits should also be considered. On top of that, it's essential to meet regulatory requirements like grid codes and power purchase agreements.

The proposed research has developed a distinctive management framework for OPF that caters to microgrids equipped with distributed energy resources. By accounting for relevant constraints and objectives and selecting an appropriate optimization algorithm, this framework will integrate the specific features and dynamics of microgrid operation to achieve enhanced efficiency, resilience, and sustainability.

Various operating conditions, such as fluctuating renewable energy availability, load patterns, and unforeseen disturbances, will be incorporated in case studies and simulations to authenticate the proposed framework's efficacy. Comparative analysis with other control methods, such as hierarchical control, distributed control, and centralized control, will be performed to ascertain its superiority. Proposed is a management framework for OPF, which will be evaluated utilizing metrics to measure its effectiveness. Indicators of performance that may be considered include the penetration rate of renewable energy, total savings in operational costs, reduced emission of fossil fuels, stability indices in the grid, and the capability of meeting load demand under different situations.

The study relies on data collected from actual microgrid installations or realistic synthetic datasets to ensure the validity and accuracy of the simulations. Additionally, software tools and simulation platforms, such as MATLAB, GAMS, or specialized microgrid simulation tools, will be used to implement and analyse the proposed OPF management framework [15]. Sensitivity analysis will be performed to investigate the impact of uncertainties, such as variations in renewable energy generation and load demand forecasting errors, on the performance of the proposed OPF system. The research will explore the robustness of the framework under different circumstances and assess its ability to adapt to dynamic microgrid environments. Figure 4. Indicates the Connection between low and high levels of renewable energy

The above approach ensures a systematic and comprehensive approach to develop and evaluate the proposed optimal power flow management system for microgrids with distributed energy resources [13,14]. It addresses the challenges of microgrid operation, considers the integration of various DERs, and aims to optimize microgrid performance while accommodating identified objectives and constraints.

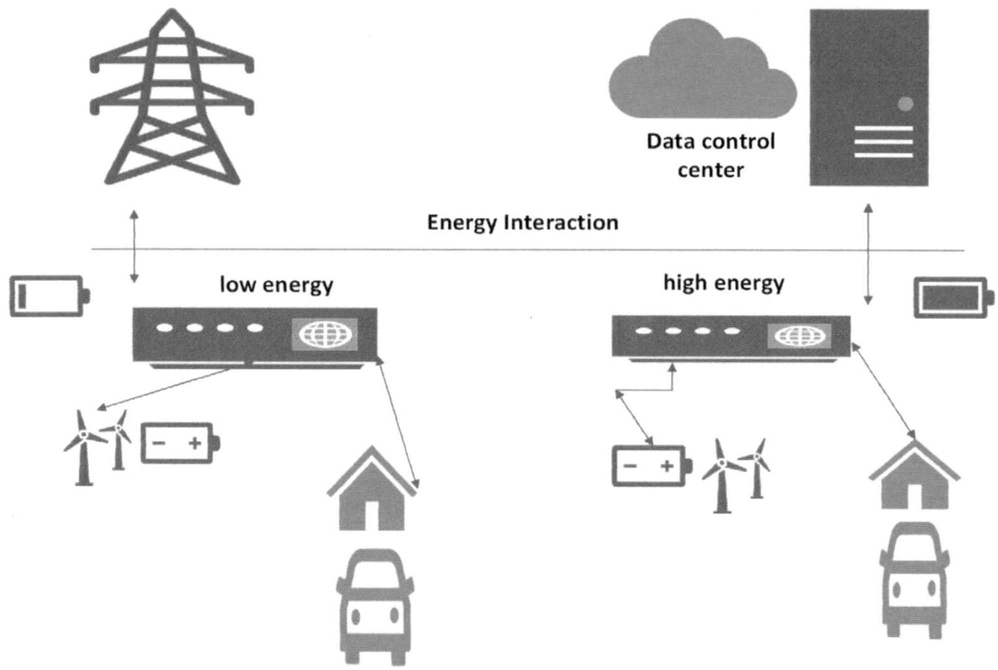

Figure 4. Connection between low and high levels of renewable energy.

5. Proposed optimal power flow management framework

The proposed optimal power flow management framework for microgrids with distributed energy resources was implemented and tested through various case studies and simulations [16]. The results obtained from the simulations were analysed and compared with conventional control methods, revealing the system's effectiveness in enhancing microgrid operation. The following are the key findings and discussions based on the conducted experiments.

5.1. Contextual investigation Situations

- Minimizing Operational Expenses: In this scenario, the objective was to minimize the operational expenses of the microgrid while satisfying the load demand and considering the availability of renewable energy sources. The proposed OPF management framework demonstrated superior cost reduction compared to conventional centralized and distributed control systems [16]. The framework effectively utilized the available renewable energy resources and energy storage systems to optimize the dispatch of generators and loads, resulting in reduced overall operational costs.
- Maximizing Renewable Energy Utilization: The goal of this scenario was to maximize the utilization of renewable energy sources in the microgrid while maintaining power balance and system stability. The proposed system outperformed conventional control methods in terms of renewable energy integration. By strategically coordinating the output of solar PV systems and wind turbines with energy storage systems, the framework achieved a significantly higher percentage of renewable energy penetration in the microgrid's overall energy mix.
- Load Variability and Flexibility: In this scenario, the microgrid was subjected to varying load patterns and unexpected disturbances to assess the framework's flexibility. The results demonstrated that the proposed OPF management system quickly adapted to changes in load demand and efficiently utilized available DERs to ensure continuous power supply. Furthermore, during unexpected disturbances, the system exhibited rapid response times and enhanced grid stability compared to traditional control methodologies.

5.2. Execution assessment measurements

- Renewable Energy Penetration: The proposed system consistently achieved higher renewable energy penetration compared to traditional control methodologies. The ability to efficiently harness renewable energy sources contributed to reducing the microgrid's carbon footprint and promoting sustainable energy practices. Figure 5 demonstrate the general Prediction for solar irradiance and solar power systems
- Operational Cost Savings: The system demonstrated significant operational cost savings by optimizing the power flow and reducing reliance on conventional fossil fuel-based generators. The efficient utilization of distributed energy resources led to reduced fuel consumption and operational expenses. Figure 6 indicates average monthly cost of electricity with DERs systems.
- Reduction in Fossil Fuel Emissions: By maximizing renewable energy utilization and reducing dependence on carbon-intensive generation sources, the proposed OPF management framework effectively decreased fossil fuel emissions, supporting environmental goals and climate change mitigation efforts. Figure 7 demonstrates General Prediction for solar irradiance and solar power systems.
- Grid Resilience and Reliability: The framework's ability to maintain grid stability during varying load conditions and disturbances highlighted its robustness. The distributed control methods in the framework facilitated faster response times and improved grid reliability.

5.3. Awareness investigation and heartiness

The sensitivity analysis revealed that the proposed system demonstrated strong performance even under uncertainties and fluctuations in renewable energy generation and load forecasting. The OPF management framework adapted quickly to changing conditions, ensuring the microgrid's stable and reliable operation. The sensitivity analysis also emphasized the system's flexibility to accommodate changes in DER output and load demand.

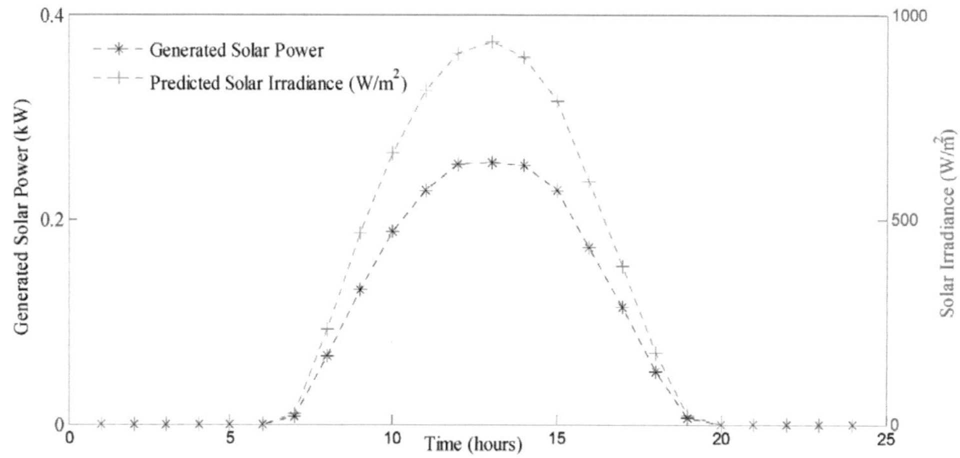

Figure 5. General prediction for solar irradiance and solar power systems.

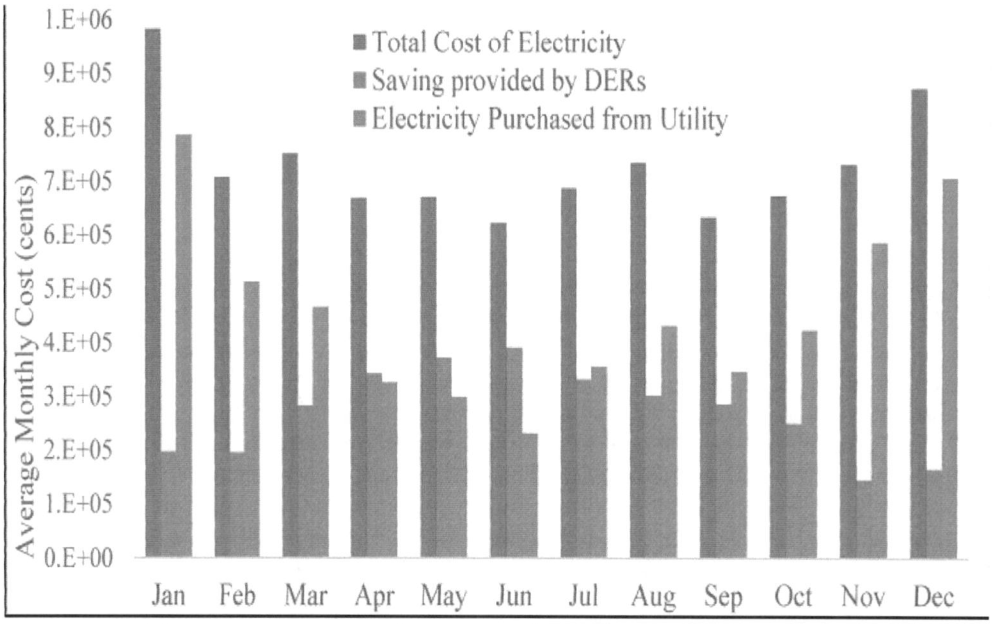

Figure 6. Average monthly cost of electricity with DERs systems.

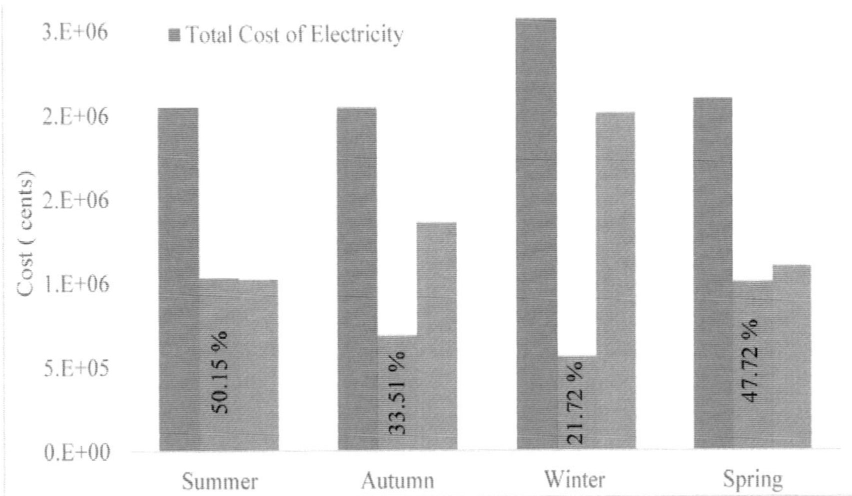

Figure 7. General prediction for solar irradiance and solar power systems.

5.4. Near investigation

The comparative study demonstrated the superiority of the proposed OPF management system over traditional control methods. The framework achieved better optimization results, improved renewable energy integration, and enhanced overall microgrid performance. The centralized control system lacked the flexibility to handle dynamic changes effectively, while the distributed control approach faced challenges in efficiently coordinating DERs. Figure 8 and figure 9 shows the output of Active power flow & Reactive power flow of the microgrid.

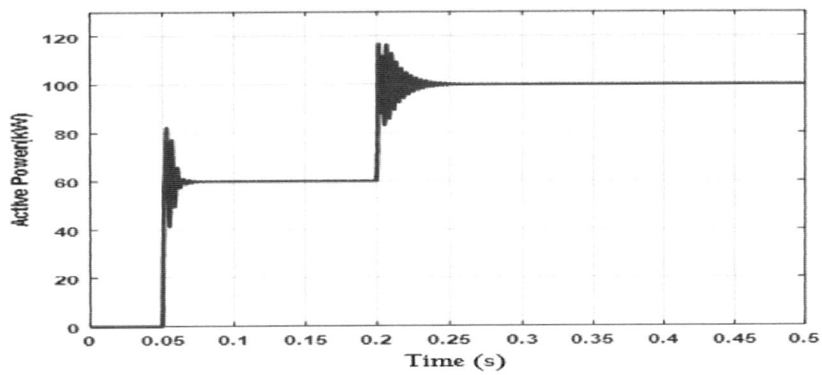

Figure 8. Active power flow.

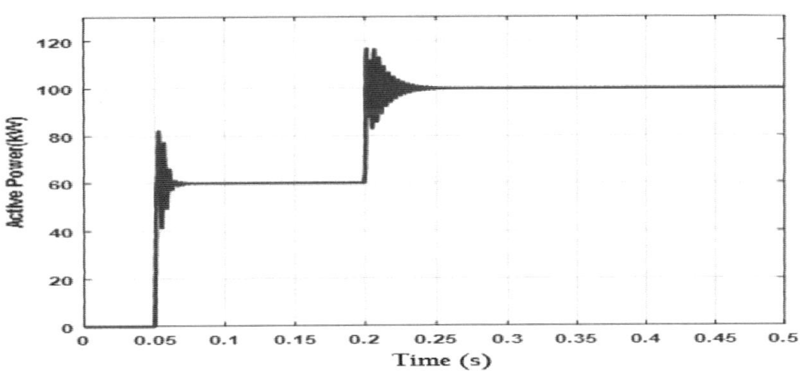

Figure 9. Reactive power flow.

IOP Conf. Series: Materials Science and Engineering 1295 (2023) 012017 doi:10.1088/1757-899X/1295/1/012017

5.5. Constraints and future exploration

While the proposed OPF management system displayed promising results, several limitations were identified during the research:

- Scalability: The system's performance for larger and more complex microgrids should be further examined to ensure its scalability.
- Real-Time Implementation: Advancing the framework to real-time implementation requires consideration of communication delays, system latency, and computational limitations.
- Market and Policy Considerations: Integrating market strategies and policy regulations into the development framework can add further complexity but is essential for sustainable deployment.

The outcomes and conversations demonstrate that the proposed OPF board structure offers a comprehensive and effective solution for optimal power flow management in microgrids with distributed energy resources. The structure exhibited better cost optimization, increased renewable energy utilization, enhanced grid stability, and reduced fossil fuel emissions. The findings provide valuable insights into the field of microgrid research and can serve as a foundation for the development and implementation of advanced control techniques in future energy systems.

6. Results and discussion

The proposed optimal power flow management framework for microgrids with distributed energy resources was implemented and tested through various case studies and simulations [16]. The results obtained from the simulations were analysed and compared with conventional control methods, revealing the system's effectiveness in enhancing microgrid operations.

In this context, the objective was to minimize the operational costs of the microgrid while meeting the load demand and considering the availability of renewable energy sources. The proposed OPF management framework demonstrated superior cost optimization compared to conventional centralized and distributed control systems [17]. The framework efficiently utilized the available renewable energy resources and energy storage systems to optimize the dispatch of generators and loads, resulting in reduced overall operational costs.

In another scenario, the goal was to maximize the utilization of renewable energy sources in the microgrid while maintaining power balance and system security. The proposed system outperformed conventional control methods in terms of renewable energy integration. By intelligently coordinating the output of solar PV systems and wind turbines with energy storage systems, the framework achieved a significantly higher percentage of renewable energy penetration in the microgrid's overall energy mix. The microgrid was subjected to varying load patterns and unexpected disturbances to assess the system's resilience.

The results demonstrated that the proposed OPF management system quickly adapted to changes in load demand and efficiently utilized available DERs to ensure uninterrupted power supply. Moreover, during unexpected disturbances, the system exhibited rapid response times and enhanced grid stability compared to traditional control methodologies.

Consistently, the proposed system achieved higher renewable energy penetration compared to conventional control methods. Efficiently harnessing renewable energy sources contributed to reducing the microgrid's carbon footprint and promoting sustainable energy practices. The system demonstrated significant operational cost savings by optimizing power flow and reducing reliance on conventional fossil fuel-based generators. The efficient use of distributed energy resources led to reduced fuel consumption and operational expenses.

By maximizing renewable energy utilization and decreasing dependence on carbon-intensive generation sources, the proposed OPF management framework effectively reduced fossil fuel emissions, supporting environmental goals and climate change mitigation efforts. The system's ability to maintain grid stability during varying load conditions and disturbances highlighted its robustness. The distributed control methods in the framework facilitated faster response times and improved grid resilience.

The sensitivity analysis revealed that the proposed system exhibited robust performance even under uncertainties and variations in renewable energy generation and load forecasting. The OPF management system adapted rapidly to changing conditions, ensuring the microgrid's consistent and reliable

operation. The sensitivity analysis also emphasized the system's flexibility in accommodating changes in DER output and load demand.
The comparison study showed how much better the proposed OPF management system was than conventional control methods. The approach boosted the integration of renewable energy sources, improved optimisation outcomes, and enhanced microgrid performance. Both the distributed control method and the centralised control system have difficulties properly coordinating DERs due to their lack of flexibility in addressing dynamic changes.
Further study is required, considering communication delays, system latency, and computing limits in real-time performance, to assure the system's scalability for bigger and more complicated microgrids. Although it might increase complexity, including regulatory restrictions and market procedures into the development framework is crucial for successful deployment. These restrictions may be addressed, and the system's capabilities can be increased, through future research. Further chances to improve microgrid operation and administration may arise from investigating the integration of cutting-edge technology, such as blockchain-based energy trading (figure 10).
In conclusion, the results and discussions demonstrate that the proposed OPF management framework offers a comprehensive and efficient solution for optimal power flow management in microgrids with distributed energy resources. The framework showed superior cost optimization, increased renewable energy utilization, improved grid stability, and reduced fossil fuel emissions. The findings provide valuable insights into the field of microgrid research and can serve as a foundation for the development and implementation of advanced control methods in future energy systems.

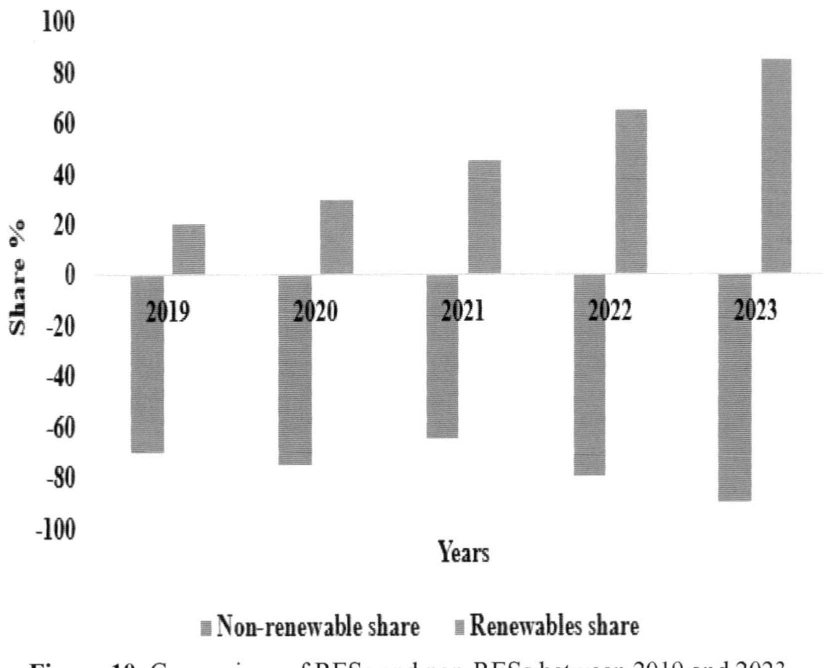

Figure 10. Comparison of RESs and non-RESs between 2019 and 2023.

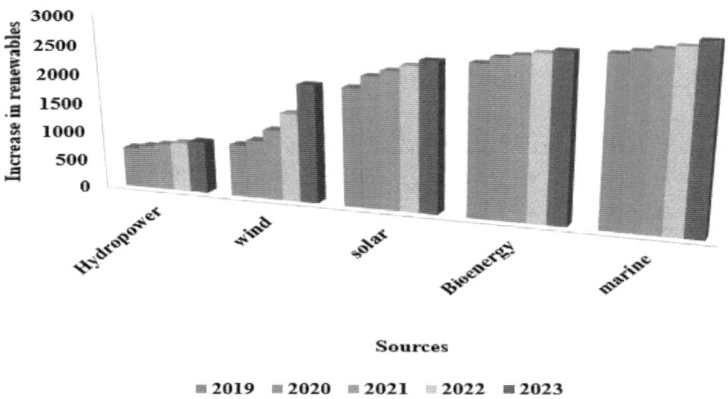

Figure 11. Global accumulated capacity of RESs from 2019–2023.

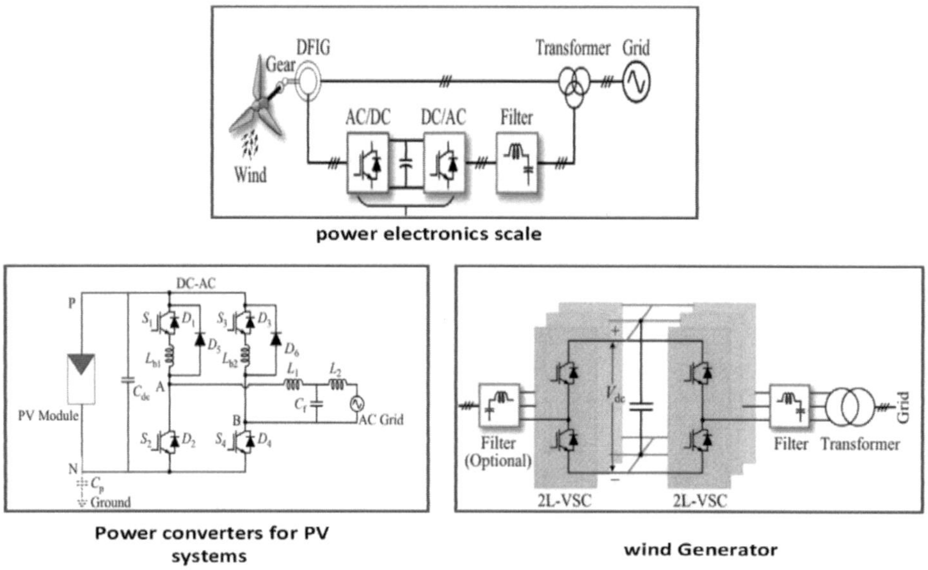

Figure 12. Control structure for grid connected PV systems: power electronics scale, power converters and wind generators.

7. Conclusion

The research work investigated the utilization of an optimal power flow management system in microgrids with distributed energy resources. The proposed system exhibited superior performance in terms of cost optimization, renewable energy integration, grid stability, and reduction in fossil fuel emissions. Through rigorous case studies and simulations, the structure demonstrated its robustness and flexibility under varying operating conditions. The results of figure 11 and figure 12 affirm the significance of efficient power flow management in achieving sustainable and reliable microgrid operation. Further advancements and research in this field can pave the way for the widespread adoption of microgrids as a viable solution for the evolving energy landscape.

8. References

[1] Hosseinnezhad V, Rafiee M, Ahmadian M and Siano P 2016 Optimal day-ahead operational planning of microgrids *Energy Convers Manag* **126** 142

[2] Ma J, Wu S and Raad E A 2023 Renewable source uncertainties effects in multi-carrier microgrids based on an intelligent algorithm *Energy* **265** 126098

[3] Lestari W 2019 Pengaruh Pelayanan Promosi dab Syariah Terhadap Minat Nasabah Dalam Memilih Asuransi Syariah (Studi pada PT.Asuransi Takaful Keluarga Cabang Palembang) *J Chem Inf Model* **2** 5

[4] Brogi A, Forti S, Ibrahim A, Kim S S, Gupta H, Vahid Dastjerdi A, Ghosh S K, Buyya R, D'Angelo G, Ferretti S, Ghini V, AbdElhalim E, Obayya M, Kishk S, Chiariotti F, Condoluci M, Mahmoodi T, Zanella A, Anglano C, Canonico M, Castagno P, Guazzone M, Sereno M, Orsini G, Bade D, Lamersdorf W, Sahinoglu M, Alouache L, Nguyen N, Aliouat M, Chelouah R, Shukla A, Chaturvedi S, Simmhan Y, Shojafar M, Pooranian Z, Sookhak M, Buyya R, Khazaei M, Mozayani N, Akbar Neghabi A, Jafari Navimipour N, Hosseinzadeh M, Rezaee A, Gil Pérez M, Huertas Celdrán A, Giardina P G, Bernini G, Pizzimenti S, García Clemente F J, Martínez Perez G, Festa G, Paglianti F, Heilig L, Lalla-Ruiz E, Voß S, Buyya R, Kathiravelu P, Van Roy P, Veiga L, Gaber M M, Aneiba A, Basurra S, Batty O, Elmisery A M, Kovalchuk Y, Rehman M H U, Zhang G A, Gu J Y, Bao Z H, Xu C, Zhang S B, Kantola R, Kabir H, Loiseau P, Chauhan M A, Babar M A, Benatallah B, Vögler M, Schleicher J M, Inzinger C, Dustdar S, Martin J P, Kandasamy A, Chandrasekaran K, Joseph C T, Baktir A C, Sonmez C, Ersoy C, Ozgovde A, Varghese B, Okafor K C, Ononiwu G C, Goundar S, Chijindu V C, Udeze C C, Zhou L L, Zhang L, et al 2018 Survey High-Performance Modelling and Simulation for Selected Results of the COST Action IC1406 cHiPSet *Future Generation Computer Systems* **29** 1

[5] Egbue O, Naidu D and Peterson P 2017 The role of microgrids in enhancing macrogrid resilience *2016 International Conference on Smart Grid and Clean Energy Technologies, ICSGCE* p 16725814

[6] Zhang S, Pei W, Yang Y, Xiao H and Kong L 2019 Optimization and Analysis of Multi-Microgrids Integration and Aggregation Operation Based on Flexible DC Interconnection *Diangong Jishu Xuebao/Transactions of China Electrotechnical Society* **34** 1025

[7] Arwa E O and Folly K A 2020 Reinforcement Learning Techniques for Optimal Power Control in Grid-Connected Microgrids: A Comprehensive Review *IEEE Access* **8** 208992

[8] Sidea D O, Tudose A M, Picioroaga I I and Bulac C 2023 Two-Stage Optimal Active-Reactive Power Coordination for Microgrids with High Renewable Sources Penetration and Electrical Vehicles Based on Improved Sine–Cosine Algorithm *Mathematics* **11** 45

[9] Farhangi H and Joos G 2019 Sample Case Studies for Real-Time Operation *Microgrid Planning and Design* Chapter 9 pp 93–116

[10] Kumar S, Yashaswini H K, Sharma N K, Bajaj M, Naithani D and Maindola M 2022 Classical Secondary Control Techniques in Microgrid Systems - A Review *Proceedings - 2022 2nd International Conference on Innovative Sustainable Computational Technologies, CISCT* p 22682768

[11] Hussain A, Bui V H and Kim H M 2019 Microgrids as a resilience resource and strategies used by microgrids for enhancing resilience *Appl Energy* **240** 56

[12] Mavromatidis G, Orehounig K, Bollinger L A, Hohmann M, Marquant J F, Miglani S, Morvaj B, Murray P, Waibel C, Wang D and Carmeliet J 2019 Ten questions concerning modeling of distributed multi-energy systems *Build Environ* **165** 106372

[13] Morvaj B, Evins R and Carmeliet J 2016 Optimization framework for distributed energy systems with integrated electrical grid constraints *Appl Energy* **171** 296

[14] Thomas D, D'Hoop G, Deblecker O, Genikomsakis K N and Ioakimidis C S 2020 An integrated tool for optimal energy scheduling and power quality improvement of a microgrid under multiple demand response schemes *Appl Energy* **260** 114314

[15] Shi W, Xie X, Chu C C and Gadh R 2015 Distributed Optimal Energy Management in Microgrids *IEEE Trans Smart Grid* **6** 1137

[16] Zia M F, Elbouchikhi E and Benbouzid M 2018 Microgrids energy management systems: A critical review on methods, solutions, and prospects *Appl Energy* **222** 1033

Design of Switched Capacitor Multilevel Inverter with Selective Harmonic Elimination

Chaitra D Buddinni[1*], and P Usha[2]

[1] Department of Electrical and Electronics Engineering, Dayananda Sagar College of Engineering, Bengaluru, Karnataka, India
[2] Department of Electrical and Electronics Engineering, Dayananda Sagar College of Engineering, Bengaluru, Karnataka, India

chaitradbuddinni@gmail.com

Abstract. In this paper, Selective Harmonic Elimination (SHE) technique is applied to control multilevel inverters that can be used to eliminate the low order dominant harmonics. This is considered a low frequency technique, in which the switching angles are predetermined based on solving a system of transcendental equations. Newton Raphson is one among the heuristic techniques utilised for selective harmonic elimination. The proposed SHE strategy is simulated along with Sinusoidal Pulse Width Modulation (SPWM) technique and the harmonic distortions are measured and compared for both techniques in MATLAB/Simulink software.

1. Introduction

Multilevel inverters have been utilised at various power and voltage levels in several commercial and utility sectors. Multilevel inverters were first conceptualised in the beginning of the 1980s.When a waveform of a voltage has more than three stages, it is said to be multilayer. Multilevel inverters' main benefit is its ability to produce reduced harmonic components at lower switching frequencies. Additionally, they produce less loss, decreased switching device blocking voltage, and less interference from electromagnetic waves (EMI). There are three primary traditional topologies for sce voltage inverters with multiple levels that are widely available. Higher voltages fav the CHB-MLI, which is additionally frequently utilised in industrial for applications requiring considerable power. Comparing the CHB-MLI architecture with the NPC-MLI and FC-MLI, it offers a dependable modularity architecture since it only needs a limited number of components [1]. Various techniques are available to regulate inverters with multiple levels [7,8]. Such methods of control are divided into two categories depending on operating frequency: small (fundamental) frequency switches along with high switching frequencies approaches. Minimal switching strategies include Space Vector Control (SVC) [2] and Selective Harmonic Elimination (SHE). On the contrary, high switching approaches employ a variety of PWM. At low frequencies, the power switches only undergo a couple of commutations every cycle. But using high-frequency methods of control, the power switch is flipped repeatedly in an interval. The SHE is a generally accepted method for inverters with multiple levels, particularly in moderate voltage and large power operations. In comparison with other regulation approaches, the SHE methodology has reduced losses during switching and reduced EMI. Additionally, it can reduce the volume of the necessary filtering on the inverter's output by eliminating the dominating lower-order harmonic. SHE precalculated the angles of switching by resolving a set of equations that are not linear. Numerous strategies and theoretical methodologies are being developed and investigated in the scientific literature to address the SHE issues associated with inverters with multiple levels [3]. All suggested solutions might be broadly categorised into three categories: repetitive procedures, consequence theory, and adaptive algorithms. Yet, using high-frequency methods of control, the power switching is flipped multiple times in an interval. The SHE is a frequently used control method for inverters with multiple levels, particularly in moderate voltage and large power operations [4]. In comparison with various control approaches, the SHE methodology has reduced losses during switching and reduced EMI. Additionally, it can reduce the size of the necessary filter on the inverter's output by eliminating the dominating lowest-order harmonic. SHE precalculated the angles at which switching occurs by resolving a set of equations that are nonlinear [5–8]. Numerous strategies and analytical methodologies are developed and investigated in the scientific literature to address the SHE issues associated with inverters with multiple levels. All suggested solutions may be broadly grouped into three categories: repetitive procedures, consequence theory, and adaptive algorithms [9–11]. Newton Raphson (NR) is a highly popular recursive method. NR was previously used to address the SHE issues in multiple levels. The technique is simple to use and effectively gets rid of the distortions. In this, nine level switched capacitor multi-level inverter is designed with selective harmonic elimination techniques. Generally, the harmonic contents can be reduced by increasing the number of levels in multi-level inverters but it results in increasing the components, size along with cost. Hence, SHE strategies are formulated for reduction in harmonics by concentrating on specific harmonic order (h=3,5,7, etc,). Here the Newton Raphson algorithm is used to generate the switching angles of Nine level SCMLI for selective harmonic elimination and compared with conventional SPWM switching technique.

2. Nine level SC_MLI

Fig.1 shows a schematic of the 9-Level Inverter's layout. Stage A with a Switched Capacitor (SC) configuration (S1-S2) and Stage B with an H Bridge cell. These two stages serve distinct roles in the operation of the inverter. In Stage A, the SC configuration (S1-S2) is used to create the desired number of voltage levels (in this case, 9) in the inverter. This configuration helps generate the multi-level AC output.There are two direct connections between Stage A SC (S1, S2) and the DC power source. Stage B, consisting of the H Bridge cell, plays a role in reversing the polarity of the inverter.

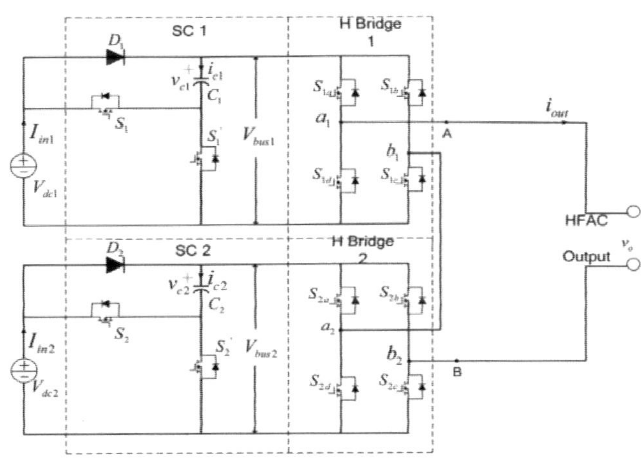

Figure 1. Nine-level inverter circuit topology

(N1= 2, N2 = 2).

The provided description outlines the operating modes of a 9-Level Inverter and how these modes are implemented through the configuration of switches in both Stage A and Stage B.

Mode 0: In this mode, switches S1a, S1b, S2a, and S2b are set to "1," and switches S1c, S1d, S2c, and S2d are set to "O." This configuration is used to charge capacitors C1 and C2 to the input voltage Vin when the output voltage is zero.

Mode 1: Mode 1 is the second requirement for mode 0. In this mode, both Bus 1 and Bus 2 have voltages of Vin. Mode 1 is achieved by closing switches S1' and S2' in Stage A and switches Sla, S1b, S2a, and S2c in Stage B. This configuration charges capacitors C1 and C2, resulting in an output voltage Vo equal to Vin.
Mode 2: To enter Mode 2, switches S1' and S2' in Stage A are turned off, and in Stage B, switches S1a, S1c, S2a, and S2c are turned on. This configuration ensures that capacitors C1 and C2 remain charged. In the second stage B, all switches S2c are closed, set to the logical value of 1. As a result, the output voltageVo equals 2 times Vin.
Mode 3: In this mode, switches S1' and S2' in Stage A are set to "1," and switches S1a, S1c, S2a, and S2c in Stage B are set to "1" or closed. During this mode, the second capacitor, C2, begins to discharge orleak. As a result of this configuration and the discharge of C2, the resultant output voltage Vo equals 3 times the input voltage Vin, which is represented as Vo = 3Vin.
Mode 4: In mode 4, switches S1 and S2 in Stage A, and switches S1a, S1c, S2a, and S2c in Stage B areset to "1" or closed. In this mode, capacitors C1 and C2 start discharging. The calculation demonstrates that the output voltage (4Vo) is twice the input voltage (2Vin).
Mode 5: For Mode 5, close switches Sl' and S2' in Stage A and S1c, Sld, S2b, and S2d in StageB. By closing these switches, capacitors Cl and C2 begin to charge, resulting in an output voltage Vo of -Vin.
Mode 6: To enter Mode 6, close switches S1', S2' in Stage A, and S1b, S1d, S2b, and S2d in Stage B. By keeping these switches closed, capacitors Cl and C2 continue to charge, and the output voltage Vo is calculated to be -2 times Vo minus 2 times Vin.

2.1. Total harmonic distortion

A crucial phase is to use the complete harmonic distortion to determine the harmonies order present in either the voltage or the current output. It can additionally be utilised to measure the amount of current or voltage produced from an electrical circuit's results. Detecting non-sinusoidal pattern requirements can additionally be done using the technique of total harmonic distortion (THD). THD, an indicator of harmonic distortion, measures the ratio of the RMS measurements of the various harmonic component (Voh) relative to the RMS data of the basic component (Vor). The switching table is provided in table 1.

Table 1. Switching table of nine level SCMLI.

ON-state switches	Output voltage	Capacitor state
S1a, S1c, S2a, S2c, S1, S2	4Vin	C1, C2Discharging
S1a, S1c, S2a, S2c, S1', S2	3Vin	C2 Discharging
S1a, S1c, S2a, S2c, S1', S2'	2Vin	C1, C2 charging
S1a, S1b, S2a, S2c, S1', S2'	Vin	C1, C2 charging
S1a, S1b, S2a, S2b, S1', S2' Or S1c, S1d, S2c, S2d	0	C1, C2 charging
S1c, S1d, S2b, S2d, S1', S2'	-Vin	C1, C2 charging
S1b, S1d, S2b, S2d, S1', S2'	-2Vin	C1, C2 charging
S1b, S1d, S2b, S2d, S1', S2	-3Vin	C1, C2Discharging
S1b, S1d, S2b, S2d, S1, S2	-4Vin	C1, C2Discharging

3. Selective harmonic elimination

In multilayer inverters, a variety of control methods are used to regulate the voltage at the output pattern. Such methods of regulation are categorised mostly according to frequency of switches. There are essentially two types of frequency- switching strategies: a) lower (fundamental) switching speeds and

b) high-frequency switching frequencies. The lowest-frequency approaches include Space Vector Control (SVC) and Selective Harmonic Elimination (SHE). As a form of high-frequency switching, especially however, several PWM, or pulse width modulation, techniques are used. The functioning power switch can only be switched once or twice every cycle when the switching rate is minimal. With high-frequency approaches, the power switching is, nevertheless, flipped multiple times in a single cycle. SHE is used to regulate a multilayer inverter, and because of an outcome of minimal switching, loss during switching and Noise are reduced. Additionally, the primary lowest-order harmonics might be chosen for elimination. Consequently, the filter used for output just must have the smallest size possible. SHE is frequently used in HVDC systems. The switching orientation is pre-calculated in SHE. The angles remove the predominant lower-order chosen harmonics while offering the basic voltage-output signal. Because of such, the THD is reduced. It is regarded considered an open-loop supervision approach since the angles for switching are computed beforehand off-line. The voltage levels shape may be described as a combination of sine and cosine signals that repeat plus an even number by using Fier's extension. There are odd and even harmonic interference in the signal. The even number of harmonics and dc offset are negated by the waveform's transverse periodicity. Therefore, just the odd harmonic interference is considered. Every triple harmonic distortion in symmetrical three-phase structures remain "0". The generated voltage pattern may often be expressed as:
Where (S) is the number of H-bride cells of the inverter.
All the switching angles are less than 90°, and are all in ascending order.

$$\boldsymbol{\theta 1 < \theta 2 < \theta 3 < 90°} \text{............ (1)}$$

The ratio of the fundamental voltage (V1) to the highest achievable fundamental voltage (V1max=n*Vdc where n is the number of levels) is known as the modulation index

$$m = \frac{\pi V_1}{4 * 3V_{dc}} =< 3m = \frac{\pi V_1}{4V_{dc}} => 3m = \frac{\pi V_1}{4V_{dc}} \text{........ (2)}$$

The SHE equation can be obtained in terms of modulation index as follows:

$$\cos(\alpha_1) + \cos(\alpha_2) + \cos(\alpha_3) = 3m \text{......... (3)}$$

$$\cos(3(\alpha_1)) + \cos(3(\alpha_2)) + \cos(3(\alpha_3)) = 0 \text{.... (4)}$$

$$\cos(5(\alpha_1)) + \cos(5(\alpha_2)) + \cos(5(\alpha_3)) = 0 \text{...... (5)}$$

$$\cos(7(\alpha_1)) + \cos(7(\alpha_2)) + \cos(7(\alpha_3)) = 0 \text{...... (6)}$$

This equation solution is very non-linear. Transcend dynamics and SHE calculations are other names for them. There are several strategies that may be used to solve this system. A sequential and adaptive algorithms are mostly often employed within the study publications. For this issue, recursive techniques like Newton-Raphson (NR) were used frequently. The main problem is that getting at the answer gets harder as the inverter's levels rises. It also needs accurate first guesses at switching angle parameters. However, adaptive algorithms handle the problem-solving differently. The primary concept is to turn the SHE issue into an optimisation task. The collection of transcendent solutions will serve as the optimization's restrictions.

3.1. Newton Raphson (NR)

Numerous iterative approaches are being employed to locate the origins of equations that are nonlinear and solve them. The Newton-Raphson approach is one of the repetitive strategies that is most frequently used and effective in addressing the issue of transcendent solutions. NR uses Taylor's periodic extension to approximate a nonlinear function over a starting point and depends on linear approach. The procedure needs an initial prediction (guess) for the value of the root x0 to determine the roots of a nonlinear function like f(x). It is important to calculate this starting estimation with the greatest caution. The most effective NR effectiveness will be when the first estimated value of x0 is closest to the actual root. On the other side, if the initial prediction was incorrect, it could be challenging to come together to NR. x0 is used as the starting point for the first repetition. Implementing the equation that follows increases the accuracy of the first guess if it fails to lead to a workable result shown in equation (1).

$$x_1 = x_0 - \frac{f(x_0)}{f^1(x_0)} \qquad (7)$$

The iterations continue until the stopping criterion of the algorithm satisfied. The estimated value is further improved by the following relation after each iteration:

$$x_{n+1} = x_n - \frac{f(x_n)}{f^1(x_n)} \qquad (8)$$

The SHE issue in multilayer inverters is addressed by the commonly utilised NR approach. The primary disadvantages of this method include: • Preliminary prediction requires practise, is difficult, and makes a big impact on the effectiveness of the algorithm.

• For certain index of modulation (M) numbers, it fails to converge. As a result, it is unable to solve the problem throughout the whole range of (M).

Implementing NR to a SHE issue is simple. The steps that follow list and describe the programme:

STEP_1: Guess initial values of switching angles

$$\theta^0 = [\,\theta_1\ \theta_2\ \theta_3\,]\ldots\ldots..\ (9)$$

STEP_2: Evaluate the value of non-linear function matrix at initial guess

$$F^0(\theta) = \begin{bmatrix} \cos\theta_1 + \cos\theta_2 + \cos\theta_3 \\ \cos 5\theta_1 + \cos 5\theta_2 + \cos 5\theta_3 \\ \cos 7\theta_1 + \cos 7\theta_2 + \cos 7\theta_3 \end{bmatrix}\ldots\ldots.\ (10)$$

STEP_3: Find the value of the Jacobian matrix using the current guess values.

$$\frac{dF^0(\theta)}{d\theta} = \begin{bmatrix} -\sin(\theta_1) & -\sin(\theta_2) & -\sin(\theta_3) \\ -5\sin(\theta_1) & -5\sin(\theta_2) & -5\sin(\theta_3) \\ -7\sin(\theta_1) & -7\sin(\theta_2) & -7\sin(\theta_3) \end{bmatrix}\ldots\ldots..\ (11)$$

STEP_4: Solve for, the value of change in the switching angles

$$d\theta^0 = INV\left[\frac{dF(\theta)}{d\theta}\right]^0 X\,(B - F^0)\ldots\ldots\ (12)$$

Where B represent the values of required harmonic amplitudes

$$B = [3M\quad 0\quad 0]^T\ldots\ldots.\ (13)$$

STEP_5: Calculate the new switching angles for the next iteration and update their values

$$\theta^1 = \theta^0 + d\theta^0\ldots\ldots\ (14)$$

STEP_6: When the greatest change in switching degrees is lower to the required error, the procedure continues from steps2 to 5. The NR must establish a requirement to terminate the method while analysing a collection of nonlinear computations, similar to any other repetitive approaches. If there is no answer, the procedure will continue running. The fulfilment of any of each of the preceding requirements have been established as the halting criteria for NR in this problem:

|n+1| is suitably small and below the desired tolerance.

- The highest number of repetitions has been reached; 100 iterations are taken into account

Following is an organisational structure (shown in figure 2) to implement the Newton-Raphson (NR) approach to the issue of selective harmonic elimination.

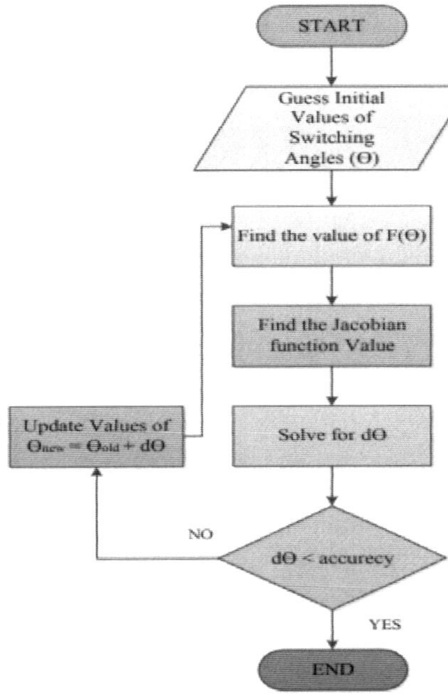

Figure 2. Flowchart for SHE using Newton Raphson method.

4. Results and discussion

The simulation parameters are provided in the following table 2

Table 2. Simulation parameters.

Parameters	Values
Input Voltage	100 V
Load power	3KW
Frequency	50 HZ
Filter Inductor	12mH
Filter Capacitor	3000μF
Coupling Capacitors	C1=0.3mF C2=0.56mF
Load Resistance	50 ohms
Load Inductance	1 mH

In figure 3 the MATLAB/SIMULINK model of MC SHEPWM is provided, and in figure.4 the control block of MC SHEPWM is provided.

IOP Conf. Series: Materials Science and Engineering 1295 (2023) 012018 doi:10.1088/1757-899X/1295/1/012018

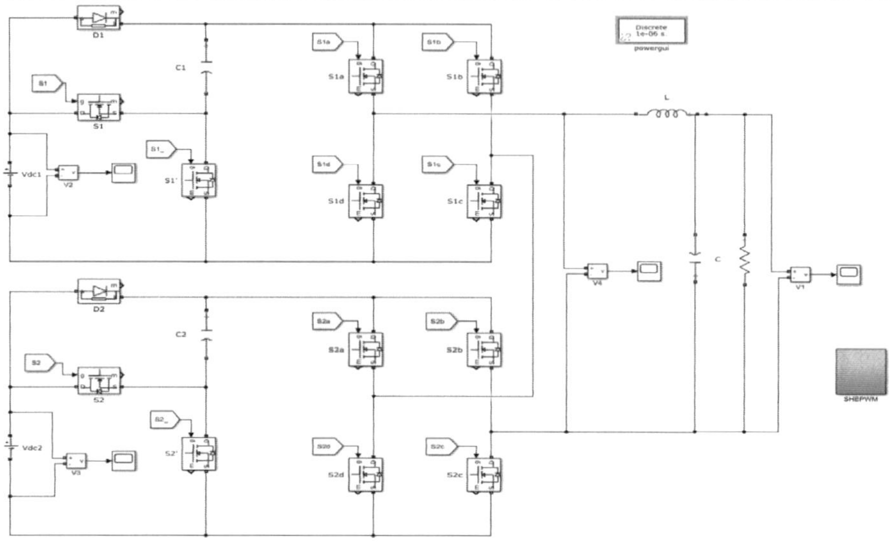

Figure 3. Simulation diagram of proposed MLI.

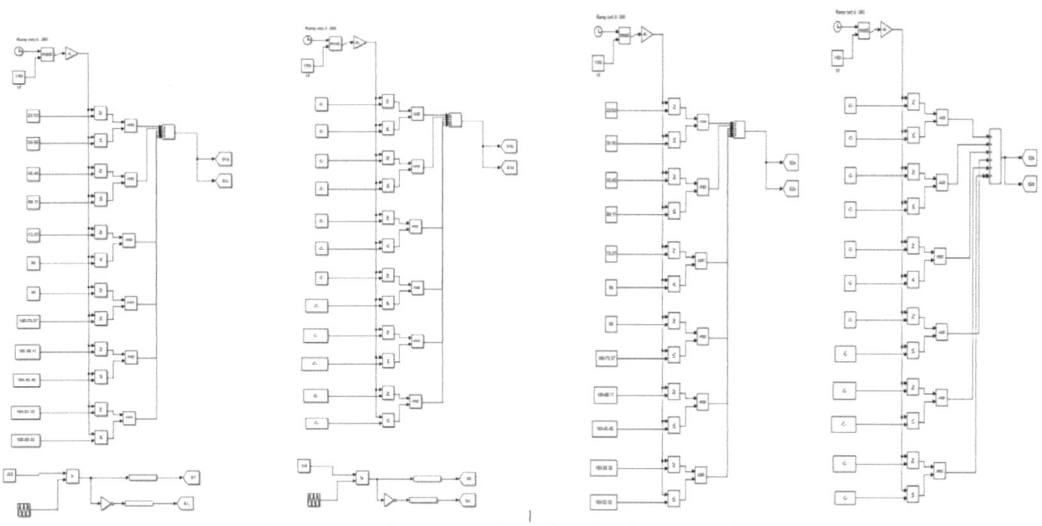

Figure 4. Pulse generation circuit of proposed MLI.

The carrier wave amplitudes in this context are assigned voltage values ranging from 1V to 12V, with each voltage level corresponding to a specific numerical representation, for instance, 1V is represented as 1 volt, 2V as 2 volts, and so on. Additionally, a standard sinusoidal reference signal has been incorporated into the system, featuring an 8V amplitude. The proposed converter has undergone simulation using the MATLAB Simulink environment, and several critical factors have been carefully considered in the simulation process. For example, we'll make some assumptions regarding the characteristics of diodes D1 and D2. Specifically, we'll assume that the forward voltage drop across these diodes is 0.6V, and their internal on- state resistance is 50 milliohms. The load resistance is taken to be 150 ohms (R0). Additionally, we have capacitors in two modules, with the value of C1 being 300 microfarads and C2 being 560 microfarads. The output frequency (fs) of the signal is set at 25 kilohertz.

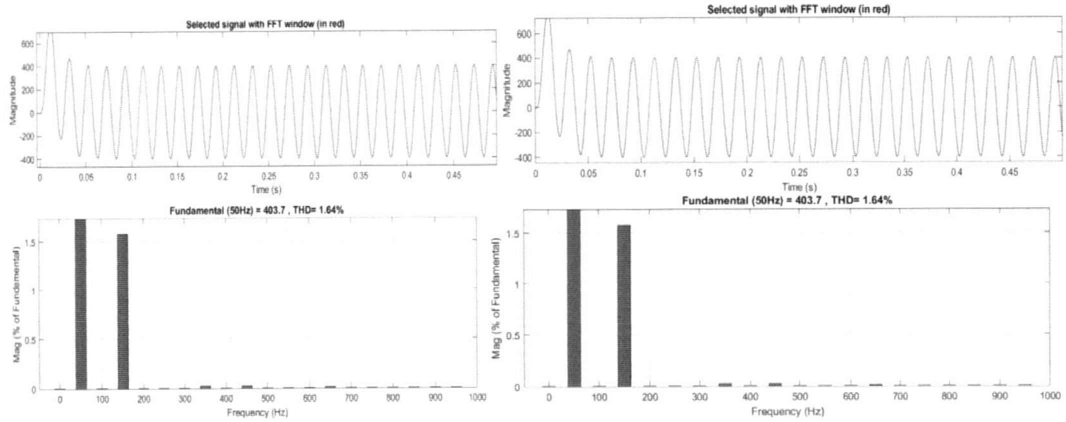

Figure 5. %THD for R(a) and RL(b) loads of proposed MLI.

The %THD of the proposed inverter is around 1.64% shown in figure 5. In this the third order harmonics (h3) is 1.54% and 5th order harmonics (h5) is 0.25% and 7th order harmonics (h7) is 0.13%. Due to the effective SHE, the %THD for both R and RL loads are less than 3%. The results of proposed inverter without filter is provided below in figure 6.

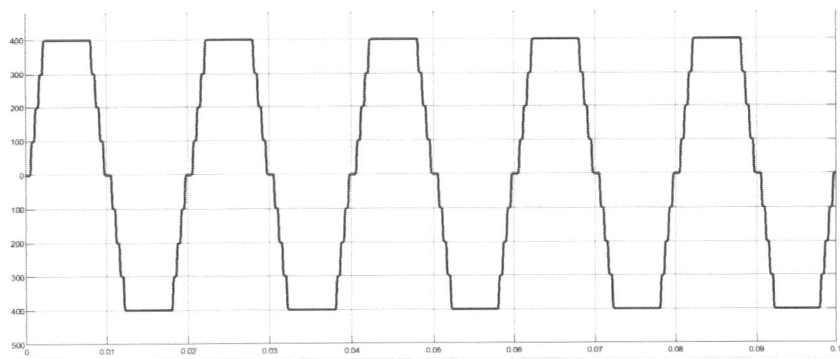

Figure 6. Output voltage without filter of proposed inverter.

In this, the positive cycle provides 4 levels and negative cycle provides 4 levels and with zero level, 9 level voltage as output of proposed MLI without filter is obtained. The MLI output voltage with filter is provided in figure 7.

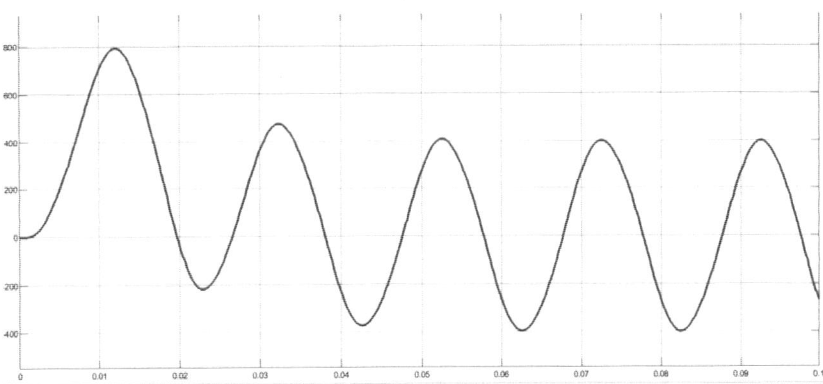

Figure 7. Typical harmonic spectra at 75% insolation.

The output of MLI is subjected to filter to get the sinusoidal voltage which is provided to load.

5. Conclusion

The research introduces an enhanced cascaded multilevel inverter (SCMLI) implemented with a switched mode power supply. The proposed circuit topology has undergone through the analysis. Compared to a conventional cascaded multilevel inverter, the count of switching devices in a nine level SCMLI can be significantly decreased. By employing the SHE method, a nine-level cascaded H- bridge MLI was constructed and subjected to analysis, revealing reduction in THD (1.64%) values.

6. References

[1] Salem A, Van Khang H, Robbersmyr K G, Norambuena M and Rodriguez J 2021 Voltage Source Multilevel Inverters with Reduced Device Count: Topological Review and Novel Comparative Factors *IEEE Trans Power Electron* **36** 2720

[2] Pan J, Yang Y, Li J, Wang F, Abdullah Y and Xu L 2020 Control of High-Performance Drive Feeding by Four-Level Hybrid Clamped Converter for Transportation Electrification *IEEE Transactions on Transportation Electrification* **6** 568

[3] Yadav A K, Gopakumar K, Krishna Raj R, Umanand L, Bhattacharya S and Jarzyna W 2019 A Hybrid 7-Level Inverter Using Low-Voltage Devices and Operation with Single DC-Link *IEEE Trans Power Electron* **34** 9844

[4] Kumar S, Yashaswini H K, Sharma N K, Bajaj M, Naithani D and Maindola M 2022 Classical Secondary Control Techniques in Microgrid Systems - A Review *Proceedings - 2022 2nd International Conference on Innovative Sustainable Computational Technologies, CISCT* p 22682768

[5] Yang Y, Pan J, Wen H, Fan M, Chen R, He L, Xie M, Norambuena M, Xu L and Rodriguez J 2021 Model Predictive Current Control with Low Complexity for Single-Phase Four-Level Hybrid-Clamped Converters *IEEE Transactions on Transportation Electrification* **7** 983

[6] Yang Y, Pan J, Wen H, Na R, Wang H, Zhang Z, Ke Z and Xu L 2020 An Optimized Model Predictive Control for Three-Phase Four-Level Hybrid-Clamped Converters *IEEE Trans Power Electron* **35** 6470

[7] Pan J, Yang Y, Cai H and Xu L 2020 Capacitor Voltage Fluctuation Minimization for Four-Level Hybrid Clamped Converter Using Improved Common-Mode Voltage Injection *IEEE Trans Power Electron* **35** 7563

[8] Pan J, Na R, Yang Y, Cai H and Xu L 2019 Capacitor Voltage Balancing and Stabilization for 4-Level Hybrid-Clamped Converter Using Selected Switching States *IEEE Trans Power Electron* **34** 12453

[9] Rasheed M, Alakkad M M A, Omar R, Sulaiman M and Halim W A 2020 Enhance the accuracy of control algorithm for multilevel inverter based on artificial neural network *Indonesian Journal of Electrical Engineering and Computer Science* **20** 3

[10] Alakkad M M A, Rasin Z, Rasheed M, Halim W A and Omar R 2020 Real-time switching thirteen-level modified CHB-multilevel inverter using artificial neural network technique based on selective harmonic elimination *Indonesian Journal of Electrical Engineering and Computer Science* **20** 3

[11] Said N A M, Saleh W A A and Halim W A 2020 Voltage harmonics reduction in single phase 9-level transistor clamped H-bridge inverter using nearest level control method *Indonesian Journal of Electrical Engineering and Computer Science* **20** 3

AUTHOR INDEX